Physiological and biochemical aspects
of heavy elements in our environment

Physiological and biochemical aspects of heavy elements in our environment

Proceedings of the symposium
Utrecht, the Netherlands, 9 May 1974

Edited by
J. P. W. Houtman
C. J. A. van den Hamer

1975
Delft University Press

PROF. IR. J. P. W. HOUTMAN
Scientific Director of the
Interuniversity Reactor Institute
Delft University of Technology
Mekelweg 15
Delft
The Netherlands

DR. C.J.A. VAN DEN HAMER
Head Department of Nuclear Biotechnique of the
Interuniversity Reactor Institute
Delft University of Technology
Mekelweg 15
Delft
The Netherlands

ISBN-13: 978-90-298-0700-5   e-ISBN-13: 978-94-010-1925-5
DOI: 10.1007/978-94-010-1925-5

Contents

Preface

In the last few years the interest in the behavior and possible function of trace elements in biological systems has strongly increased. This seems to be stimulated by three different factors: i.e. 1. the slowly growing notion about the importance of some essential elements in the metabolism of plants and animals, including man, 2. the increasing extent of environmental contamination caused by the disposal of heavy elements – both essential and toxic – and 3. the increasing sensitivity achieved in the analysis of these elements by modern techniques.

The impact of the last mentioned development has been strong. Enormous improvements have indeed been made in lowering the limits of sensitivity of analytical methods maintaining at the same time good precision. Though there is a strong competition between some techniques, neutron activation analysis shows special advantages for multi-element analyses in organic materials. As a result a relatively important part of the literature on trace element behavior in the biosphere, though often phenomenologic in character, can be found in journals and transactions of symposia involving radiochemistry. This phenomenon does not readily stimulate the communication and scientific exchange with nutritional, toxicological, medical or biochemical experts who study the metabolic pathways of trace elements. On the other hand, the environmentalists who have become increasingly alarmed about the concentrations of heavy elements observed by the analytical specialists in soil, water, crops, food, but also in organs of various animals, are insufficiently informed by the fundamentalists and grope in the dark in their attempts to interpret data and to weigh the dangers. As Kothny* has recently stated in the preface to a symposium: 'We must understand the metabolic process before we condemn the presence of some trace elements with apparently no value', a statement to which we wish to subscribe. In this respect special

* E.L. Kothny (ed.), 'Trace Elements in the Environment', *Amer. Chem. Soc.*, (Washington, 1973).

attention should be paid to the principles of essentiality and toxicity. By now we know that these do not exclude each other. Some elements can be both essential and toxic depending on concentration and the presence of cofactors. Also here more clarity should be sought.

It is therefore highly important that the various scientists, each working in some special area of this enormous field, come together and stimulate a more systematic cooperation between the various disciplines involved. In the Netherlands this consideration has led to the organization of a 'Meeting on the Physiological and Biochemical Aspects of the Heavy Elements in our Environment' (Utrecht, May 9, 1974) in which a survey of the various areas has been presented by a few specialists in environmental contamination, animal nutrition, animal physiology, medicine and biochemistry, of whom some have an international reputation. This meeting was initiated by the Interuniversity Reactor Institute at Delft, because of its interest in the use of neutron activation analysis and radiotracer techniques in the study of biomedical problems. It was organized under the auspices of the Genootschap ter bevordering van Natuur-, Genees- en Heelkunde (Society for the Advancement of Physics, Medicin and Surgery) at Amsterdam and was further sponsored by the Netherlands Ministry of Education and Science and by the Central Laboratory TNO. During this meeting it was felt that the five contributions formed together a more or less unique survey of present-day knowledge and could also be used by others in planning new research contributions. Therefore, the manuscripts have been prepared for publication in bookform. It is our hope that it will find its way to many scientists all over the world.

J. P. W. Houtman
C. J. A. Van den Hamer

Authors

DR. H.J. HUECK

Dr. H.J. Hueck studied biology at the universities of Leiden and Utrecht, and in 1952 he was graduated on an ecological thesis.
After having held several appointments in the industry, he entered the employment of TNO in 1952, where he now heads the Department of Biology of the Central Laboratory TNO in Delft.
The main research subjects of this department are the desirable and undesirable biodegradability of materials and the evaluation from a viewpoint of environmental toxicology of waste materials in the aquatic environment.
He is Honorary President of the 'International Biodegradation Research Group', and Vice President of the 'Biodeterioration Society'.

PROF. DR. M. KIRCHGESSNER

Prof. Dr. M. Kirchgessner, who is now Director of the Institute of Nutrition Physiology of the Techn. University of Munich, studied agriculture at the University Hohenheim and chemistry at the University Stuttgart. He finished his thesis in 1955 and his Habilitationsschrift in Nutrition Physiology in 1958. His research interests center on the metabolism of trace elements – questions like their absorption and availability in the intermediary metabolism – and of B-vitamins, but also on the energy requirements of animals and their effect on protein biosynthesis. He published 3 books and some 300 papers.
Prof. Kirchgessner was awarded the Oskar-Kellner Price of the Landwirtschaftliche Untersuchungs- und Forschungsanstalt in 1957 and the Lehmann-Henneberg Price of the University Göttingen in 1972. Since 1967 he is vice-president for Animal Nutrition of the European Association for Animal Production.

DR. J. PAŘÍZEK

Dr. Pařízek is head of the Laboratory of Physiology and Pathophysiology of Reproduction, Institute of Physiology, Prague. In the past ten years much of his work has been focussed on the damage by heavy metals – particularly by cadmium – on reproductive organs and on the influence of selenium on these effects.

PROF. DR. I. H. SCHEINBERG

After graduation from Harvard Medical School and graduate study at Harvard and the Massachusetts Institute of Technology, professor dr. I. H. Scheinberg moved to New York City. Since 1955, he has been Professor of Medicine at the Albert Einstein College of Medicine, Bronx, N.Y. Here, much of his research efforts have concentrated on the metabolism of copper and of ceruloplasmin, and on the care of patients with Wilson's disease, areas in which he is widely known. He is Chairman of the Panel on Copper of the Committee on the Medical and Biologic Effects of Environmental Pollutants of the National Research Council of the National Academy of Sciences in the United States.

DR. C. J. A. VAN DEN HAMER

Dr. Van den Hamer studied chemistry at the State University Utrecht. After completion of his thesis in 1960 ('The Carbohydrate Metabolism of the Lactic Acid Bacteria') he was connected for some years with the Clinical Chemistry Laboratory of the Willem Arntsz Stichting (Utrecht) where – apart from clinical chemistry (since 1962 as a Certified Clinical Chemist) – he did research on the relation between biochemistry and mental health. In 1964 he joined the group of prof. Scheinberg in the Dep. of Medicine of the Albert Einstein College of Medicine (Bronx, N.Y.) were he was mainly involved in various aspects of the metabolism of copper. Since 1971 dr. Van den Hamer is head of the Dep. of Nuclear Biotechnique of the Interuniversity Reactor Institute, Delft. This department, though also involved in research in the field of nuclear medicine, concentrates its efforts mainly on the physiology of trace metals.

Contamination of the environment by some elements

H. J. Hueck

1. INTRODUCTION

In a geochemical sense, all elements are natural, but they differ in their occurrence and their distribution over the earth. Relatively few of the elements are known to have a function in living systems. With due respect to George Orwell we may therefore state that 'all elements are natural, but some are more natural than others'. In the long process of genetic evolution, organisms have adapted themselves to the availability of certain elements. The essential elements in one way or another take part in metabolic processes. The other elements are either inert or toxic to living systems. Towards toxic elements, organisms have developed defence mechanisms or avoidance reactions. The result is a sensitive equilibrium between organisms and their chemical environment. In ages past, slow changes in this equilibrium have always occurred, affecting both the distribution of elements (by biogenic deposits) and the composition of the flora and fauna. Such changes have taken place over millions of years. With the advent of man, however, a much more rapid process had begun, viz. the technological use of certain elements which markedly affects the distribution of these elements in the biosphere. We may rightly suspect that such short term changes have a damaging influence on organisms if their adaptive potential is exceeded. It is difficult to assess the extent to which mankind is indeed contaminating its environment. Förstner and Müller (1974) have summarized much of the available evidence for the contamination (pollution) of the environment by heavy elements, and I should like to quote a few of their results as an illustration.

One may relate world-consumption of metals to the natural occurrence of these metals in the soil in order to arrive at an index showing their relative contamination potential (Table I).

Table I refers to potential contamination. Evidence for actual contamination can be found in considering a time series of actual concentrations of certain elements in e.g. rivers or in analysing deposits in rivers.

1

Table I. *Potential contamination of the environment due to human industrial consumption. Adapted from Förstner and Müller, 1974.*

World consumption of heavy metals (1968/1969).
Metal concentrations in soils and
index of relative pollution potential.

Metals	Metal consumption × 1,000 t/a	Soils (ppm)	Index of relative pollution potential
Fe	400,000	38,000	1
Mn	9,200	850	1
Cu	6,400	20	30
Zn	4,600	50	10
Pb	3,500	10	35
Cr	1,700	100	2
Ni	493	40	1
Mo	57	2	3
Co	19	8	0.2
Cd	15	0.06	25
Hg	10	0.03	30

Table II. Comparison of ion-composition of some rivers in about 1890 and 1970. Adapted from Förstner and Müller, 1974.

	bicarbonate (mg HCO_3^-/l)	sulphate (mg SO_4^-/l)	chloride (mg Cl^-/l)	calcium (mg Ca^{2+}/l)	magnesium (mg Mg^{2+}/l)	sodium (mg Na^+/l)

Danube

Rhine

Weser

□ 1887/1893   ▓ 1971   □ 100 mg/l   □ 10 mg/l

2

*Table III. Relative contributions of 'natural' and 'anthropogenic' con-
centrations of heavy metals in the Lower Rhine. Adapted from Förstner and
Müller, 1974.*

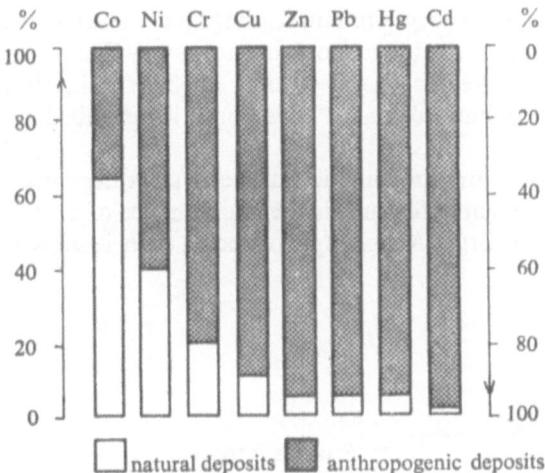

Table II shows the changes in ionic composition of some European rivers over a period of about 80 years.

Table III shows the contribution by human civilization as compared with the natural background concentrations of some heavy elements in the river Rhine.

These examples indicate that in the last hundred years considerable changes must have taken place. It is up to the 'environmental toxicologist' to assess to what extent this pollution influences the biosphere. Let us first briefly consider what we shall regard as pollution in relation to organisms.

2. ON THE NATURE OF POLLUTION

The physical existence of mankind is possible only in relation with an environment belonging to it. Between man and his physical, chemical and biological environment many relations exist, which, moreover, are inter-dependent. The entirety of mankind and its environment constitutes the human ecosystem. This system possesses self-stabilizing properties arising from negative feed-back mechanisms in many of its essential relations.It has a certain harmony of its own, which certainly is not of a static nature, but which constitutes a dynamic whole that shows continual fluctuations.

3

'Damage' to such a system therefore cannot be equated to any arbitrary structural or functional change in it. It must be something more fundamental than that. Therefore, we might define pollution as environmental damage that results from 'a process developing from the interaction of a noxious agent and the human ecosystem and leading to a disturbance of the harmonious functioning of the biocoenosis in such a way that normal variations in its functioning are surpassed' (Odum, 1963; Hueck, 1971). We must conclude that 'pollution' makes sense only if described in a quantitative way.

For maintaining its structure and function an ecosystem depends on many factors. One of the essential features is the maintenance of the well-known cycles of matter or energy. An example of such a cycle is shown in Fig. 1. (Woodwell, 1970).

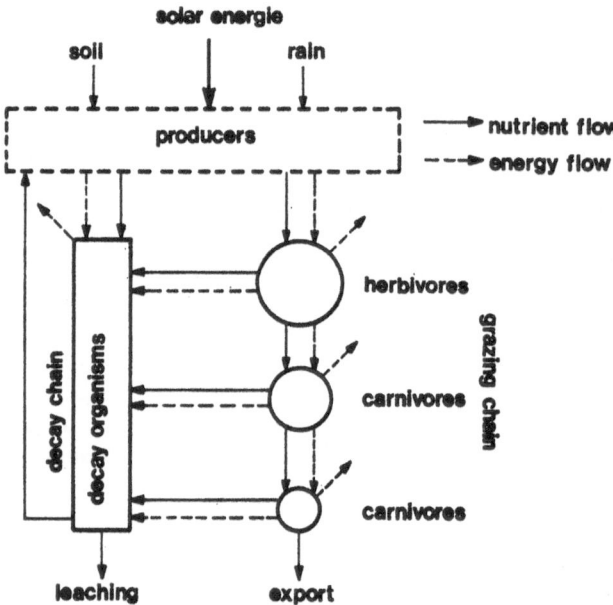

Fig. 1. Adapted from Woodwell, 1970.

Interference with this recycling of matter must be regarded as one of the severest forms of pollution.

The harmonious functioning and stability of the whole ecosystem may be thought to be essential to this as many types of organisms take part in the processes mentioned. Organisms are grouped in a series of trophic levels (Lindeman, 1942) (cf Fig. 1). These trophic levels are related to one another by an exchange of matter and energy (food). Structure and functioning of an ecosystem is intimately connected with the number and variety of organisms which it contains. The effects of pollutants should therefore be measured in terms not only of mortality, but also of growth and reproducibility of representative organisms in such a system, e.g. species belonging to different trophic levels of the system.

3. ENVIRONMENTAL TOXICOLOGY OF COPPER AND MERCURY

3.1. *Introduction*

I have stressed before that the concept of pollution makes sense only if it is considered quantitatively. The data needed for such an assessment are often completely lacking. For copper and mercury more data are at hand (De Vos, 1971; Hueck, 1972). In order to arrive at a meaningful comparison, the data will be limited to a small geographical area, viz. the Netherlands. Furthermore, toxicological data will be given as far as possible from experiments with the same organisms, and preferably done by the same experimenter. A further restriction is that only data for the aqueous environment are given. Relevant questions in relation to the environmental stress due to these metals are:
1. How much is added annually to our environment.
2. What amounts can be found in Dutch waterways and in the North Sea.
3. What amounts can be found in organisms exposed to copper and mercury.
4. What toxic symptoms are shown by representative organisms.

3.2. *Release of copper and mercury into the environment*

Estimates of the release of copper and mercury were given by Lageveen-van Kuyck (1972) and Beek (1971). Although some of their data are uncertain, these do provide an insight into the order of magnitude. Some details have been added from other sources (Hueck, 1971; Hueck, 1972; De Groot and Fonds, 1972). The data are summarized in Table IV.

The position of copper and mercury is quite different as to their burden on the environment. For mercury the main source would appear to be the industry. Leading industries in The Netherlands recently have discontinued the use of mercury e.g. as a preservative in paints or diminished the use as a catalyst so that the actual situation may be more favorable. For copper agriculture appears to be a major source of contamination even more so than industry.

3.3 *Presence of copper and mercury in water and organisms*

The following data are mainly based on review publications (De Vos, 1971; Hueck, 1972) of the Organization for Applied Scientific Research TNO in The Netherlands, Koolen (1973) and Roskam (1970). Again they refer to the Dutch situation. They are summarized in Table V and VI.

It should be added that occasionally very high local concentrations of copper have been observed (Hueck, 1972) viz. 40–145 $\mu g/l$ in some cases related to industrial pollution. The number of observations being limited, the concentration of mercury in coastal waters can only tentatively be estimated at about 0.1 $\mu g/l$. It is clear that coastal waters carry a load of copper which can definitely be attributed to pollution. For both metals there are definite indications that they act as local or occassional pollutants of our waterways. Some data on observed concentrations of both metals in organisms are summarized in Table VI. Having been obtained by

Table IV. Release of copper and mercury into the Dutch environment (data 1969–71)

		copper 10^3 kg/year			mercury 10^3 kg/year
source:			*source:*		
River-borne			*River-borne*		
Rhine		2900	Rhine		70
Other rivers	5% Rhine		Other rivers	5% Rhine	
Industry		500	*Industry*		46
Metal plating	200		Electrolytic		
Electronics	300		chlorine	20	
			Paint	24	
			Other	2	
Agriculture		975	*Agriculture*		4
Fertilizers	225				
Fodder and					
pesticides	750				
Other sources		95	*Other sources*		5
Input into the North Sea			*Input into the North Sea*		?
Rivers + landflow		ca. 3000	(100)		
Antifouling paints		60			

Table V. Mean concentrations of copper and mercury in water in the Netherlands

Type of water	copper		mercury	
	ppb	μmol/1	ppb	μmol/1
Rhine[1]	14	0.22	0.3	0.001
Meuse[1]	4	0.06	(0.2)	(0.001)
Inland waterways general	10–20	0.16–0.31	0.1–4.0	0.0005–0.020
North Sea				
coastal water	6–12	0.09–0.19	–	–
open water	1–3	0.02–0.05	0.03	0.00015

1. in aqueous phase

$$1 \text{ ppb copper} = 1 \ \mu g/1 = \frac{1}{63.5} \ \mu mol/1$$

$$1 \text{ ppm mercury} = 1 \ \mu g/1 = \frac{1}{200.6} \ \mu mol/1$$

Table VI. Mean concentrations of copper and mercury in some organisms in the Dutch environment, 1970/1971

Organisms	copper		mercury	
	ppb (dry weight)	μmol/kg	ppb (wet weight)	μmol/kg
Marine algae (Fucus)	5000–10000	80–160	70–140	0.35–0.70
Zooplankton	30000–200000	470–3100	400	2.0
Daphnia	30000	470		
Mussels				
from buoys in open sea			50	0.25
Dutch coast	7000–15000	110–240	100–200	0.50–1.0
Eems estuary			200–400	1.0–2.0
Fishes				
Freshwater			140–850	0.70–4.2
Marine	1000–7000	20–110	70–340	0.35–1.7
Seabirds			900–1400	4.5–7.0

different analytical methods, the data given in this table are difficult to compare. Whereas mercury usually is determined by neutron-activation analysis, which is based on wet weight, copper is determined by atomic absorption spectrometry, which refers to dry weight. As a rule of thumb we may assume dry weight to be about 20% of wet weight. Even with this

7

proviso, it is apparent that copper contents of organisms are generally much higher than mercury contents. Since copper is an essential element and mercury is not, this is not at first sight surprising. Considering the figures for mercury, we find that only coastal mussels are slightly over-burdened with mercury. The high mercury content of mussels from the Eems estuary probably results from industrial pollution by a local chlorine factory, which uses mercury in its manufactoring process.

Miss Adema of the Central Laboratory TNO has carried out experiments on the accumulation of copper and mercury in one of the organisms mentioned, viz. mussels. The results are shown in the Figs. 2 and 3.

These figures show that the tolerance of the organisms to copper and mercury is quite different. The essential metal copper is perhaps maintained at its natural concentration in the organism by some stabilizing mechanism. If the environmental burden exceeds the tolerance of this mechanism, the tissue-concentration rises and soon leads to mortality (range about 5). With mercury, which, as far as we know, has no essential function in organisms, tissue levels respond to a much wider range of

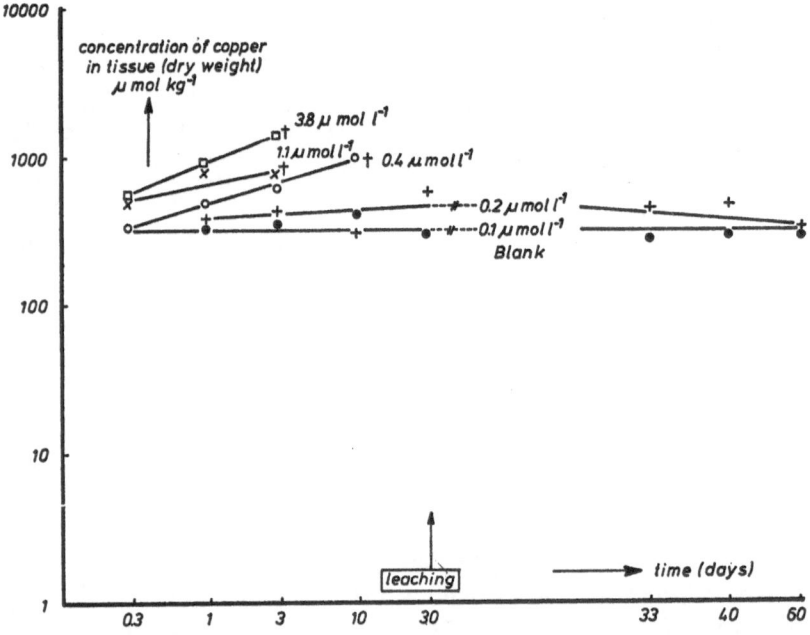

Fig. 2. Accumulation and elimination of copper in mussels at various concentrations in water.

environmental concentrations before leading to death (range about 1000). We shall return to this after having considered some toxicological data.

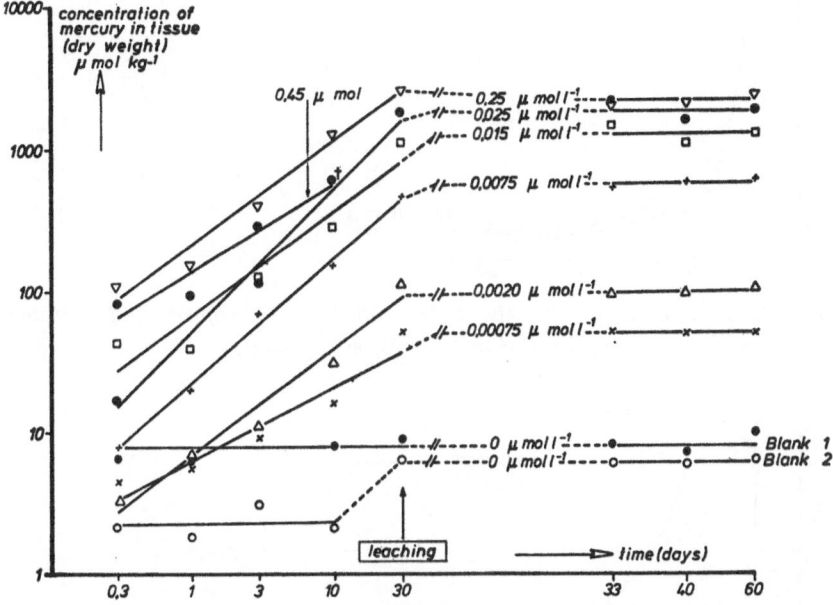

Fig. 3. *Accumulation and elimination of mercury in mussels at various concentrations in water.*

3.4. *Environmental toxicity of copper and mercury*

As discussed earlier, environmental toxicology should not be limited to considering human toxicity, but a wide variety of organisms should be taken into account. Moreover the usual data on acute toxicity are insufficient: because for persistent chemicals chronic toxicity appears to be far more important. Such data are scarce. If for want of anything better one uses acute toxicity data, one should be aware of the pitfalls in doing so. Figs. 4 and 5 give a comparison of the toxicities of copper and mercury to two different animals. The test was carried out in a constant-flow system, which allowed a rather precise estimate to be made of LC_{50} as a function of the duration of the experiment. It is quite clear that there are enormous differences between animals exposed for four days (the usual period in acute tests) and those exposed for 30 days, in what might be called subchronic experiments. Comparison of the figures reveals another relevant

9

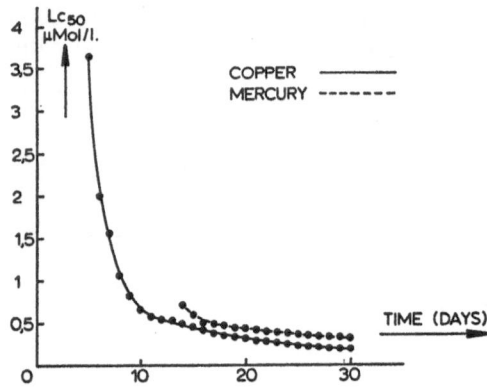

Fig. 4. Mortality of mussels under influence of copper and mercury. Expressed as LC_{50}. (Data of miss D.M.M. Adema)

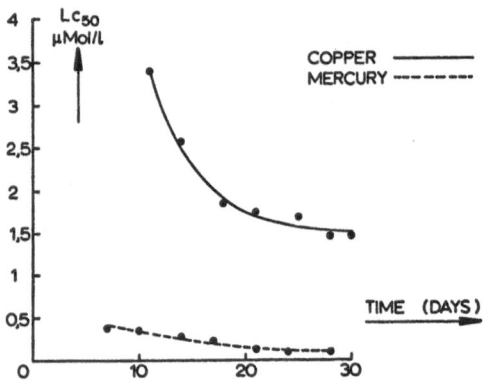

Fig. 5. Mortality of gammarids under influnece of copper and mercury. Expressed as LC_{50} (different batches of test animals). (Data of miss D.M.M. Adema)

fact, viz. that even for such seemingly simple substances as metals, great differences in sensitivity are apparent. Mussels are so sensitive to copper that to them it is as poisonous as mercury, a metal which to most other organisms is the more toxic one, as shown in the figure for Gammarids as test animals. Another point in question is that the outcome of a toxicity test depends very much on its nature, so that simple toxicity figures generally have only a very relative value. The data given in Table VII which compares the toxicities of copper and mercury to different or-

10

Table VII. Toxicity to various organisms of inorganic compounds of copper and mercury.

Organisms	Type of test	copper		mercury	
		ppb	μmol/1	ppb	μmol/1
Primary producers					
Chlorella	LC_{100}, $CuSO_4$, $HgCl_2$	10000	157	3000	15.0
–	no-effect level	1	0.02	4	0.02
Chlamydomonas	50% growth retardation	8	0.13	12	0.06
Secondary producers					
Daphnia	no-effect level	10–17	0.16–0.27	3–5	0.01–0.02
Gammarids	LC_{50}, 30 days	90	1.4	22	0.11
Mussels	LC_{50}, 30 days	15	0.24	70	0.35
Oysters	LC_{50}, 4 days, embryos	100	1.6	5.6	0.03
Consumers					
Fish	no-effect level	30–800	0.5–12.6	< 8	< 0.04
Man	fatal dose, acute toxicity in mg/kg body weight	1000–8000	15.7–126	8–330	0.04–1.6
Rat	LD_{50} mg/kg body weight	140	2.2	37	0.18
Decay organisms					
Bacteria					
'sewage organisms'	100% bacteriostasis	25000	394	2000	10.0
–	50% inhibition	–	–	600	3.0
Fungi	100% fungistasis ($CuSO_4$, $HgCl_2$)	10^5	1575	600–10000	3.0–50

LC = lethal concentration
LD = lethal dose

ganisms, should therefore be regarded with some reserve. The selection made is rather arbitrary and serves only as an example. The data were mainly drawn from the same source as those in previous tables, and as far as possible they were based on the same type of test with inorganic compounds. Organo-mercury and organo-copper compounds often are more toxic. From this table it appears that organisms show large differences in sensitivity. Clearly micro-organisms belong to the least sensitive ones, though copper and mercury are frequently used to control them. The larger aquatic organisms are far more sensitive. The last figure(s)

11

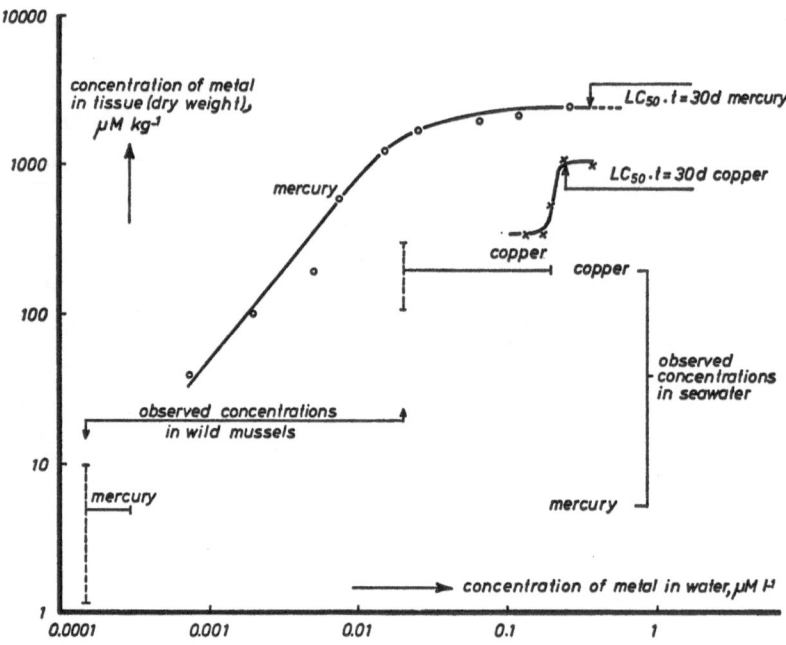

Fig. 6. Accumulation of copper and mercury in mussels after 30 days (in vitro).

constitutes an attempt to correlate, to the extent to which this is possible, the biological effects of copper and mercury with the actual concentrations in which they are found in organisms and their aquatic environment. We observe that, owing to the narrow margin of tolerance of mussels to copper, the potential danger of the latter to mussels is far greater than that of mercury. Moreover, one gains the impression that the 'natural' content of copper in mussels is rather well geared to environmental concentrations. With mercury no such mechanism is apparent. It is tolerated over quite a large range of concentrations before the onset of mortality which appears to occur after some passive 'saturation' of the organisms with mercury has taken place. The tissue level of mercury apparently is determined mainly by its concentration in the aqueous environment, and not by some internal regulating mechanism.

It must be stressed that the results shown refer to inorganic forms of copper and mercury. Methyl mercury, a product of bacterial conversion in the environment, is perhaps more dangerous to many organisms and especially man, than inorganic forms of the metal. The contamination of

12

our environment begins, however, mainly with the inorganic forms of copper and mercury. I hope to have given you an impression of the impact of this primary contamination on a small but important part of our environment.

SUMMARY

'All elements are natural, but some are more natural than others'. Of all elements only a limited number are essential to living beings. Though some, like mercury, are virtually omnipresent, they have no known function in biological processes. Apart from this qualitative aspect we have to consider quantitative aspects. Man is continuously shifting elements from one place to another, which means that in some places or at some moments 'unnatural' amounts may occur.

In this presentation we will restrict ourselves to the influence of the 'shifts' on the biosphere. We know that organisms have become adapted to their environment in an evolutionary process which has lasted millions of years. It must be expected that the short-time changes due to the industrial revolution may have some influence on the delicate balance between organisms and their environment. As 'environmental toxicology' is essentially a quantitative branch of science, it is not sufficient to point out possible harmful effects in a qualitative way as many worried environmentalists do, but we must try to assess such effects also quantitatively.

This is not an easy task as the study of ecological relations and mechanisms is still in its infancy. A limited approach to such a quantitative assessment is given for the essential element copper and the non-essential element mercury as far as they occur in the aqueous environment of the Dutch Delta. The following questions are considered:
1. how much is added annually to this part of the environment;
2. what amounts are actually found in Dutch waterways and in the North Sea;
3. what amounts can be found in organisms exposed to copper and mercury;
4. what toxic symptoms are exhibited by representative organisms.

4. REFERENCES

Adema, D. M. M., *Personal communication*, unpublished observations (1974).
Beek, W. J. C., 'Quantitative aspects of environmental pollution', in *Mens en Milieu, prioriteiten en keuze*, Kon. Inst. van Ingenieurs, (The Hague, The Netherlands 1971).

Doorgeest, T., 'Het kwikgebruik in de Nederlandse verfindustrie', *Verfkroniek* **45**, (1972) 41.

Förstner, U. and E.G. Müller, *Schwermetalle in Flüssen und Seen als Ausdruck der Umweltverschmutzung*, Springer Verlag, (Berlin, Heidelberg, New York 1974).

De Groot, A.J. and A.W. Fonds, 'Voorkomen en gedrag van zware metalen in de Nederlandse delta', *De Ingenieur* **84** (1972) G19.

Hueck, H.J., 'Biodeterioration and environmental pollution', *Int. Biodetn. Bull.* **7** (1971) 81.

Hueck, H.J., (Edit.), 'The importance of copper to our environment' *TNO-Nieuws* **27**, (1972) 415 (Dutch with English Summary).

Koolen, J.L., 'De kwaliteit van het Maaswater in Nederland', H_2O **6** (1973) 3.

Lageveen-van Kuyck, H.J., 'Applications and use of copper in the Netherlands', *TNO-Nieuws* **27**, (1972) 428.

Lindeman, R.L., 'The trophic dynamic aspects of ecology', *Ecology* **23** (1942) 399.

Odum, E.P., *'Ecology'*, Holl. Rinehart & Wilson, (New York 1963).

Roskam, P.T.H., 'De verontreiniging van de zee', *Chem. Weekbl.* **66**, (1970) 56.

De Vos, R.H., (Edit.), 'Mercury and its compounds in the Dutch environment', *TNO-Nieuws* **26**, 371 (1971) (Dutch with English summary).

Woodwell, G.M., 'The energy cycle of the biosphere', *Sci. Amer.* **223** (3), (1970) 64.

Absorption and metabolic efficiency of essential trace elements

M. Kirchgessner

Elements which do not surpass a concentration of 50 mg per kg of body weight, which function as activator or component of various enzymes and which are absolutely indispensable for the normal metabolism of cells are considered as essential trace elements. Therefore, they must be present in the food. Man and animal make use, however, of only part of the trace elements ingested with the food just as is the case with other nutrients and minerals. Only part of the trace element passes through the gastrointestinal lining into the body, it is absorbed in other words, while the remaining portion is| excreted via the feces. The specific metabolism of a trace element determines to which extent the absorbed portion is stored in various body regions or utilized for metabolic functions and syntheses in the body and to which extent it is excreted again via the feces, i.e. the endogenous portion and via the urine. The efficiency of the utilization in the intermediary metabolism is affected, however, by the chemical bondage. For illustration and definition of this concept a simplified scheme is presented in Fig. 1.

The utilization of trace elements, defined as the total availability for the organism, can be divided into component parts. In providing adequate dietary intake of essential trace elements to man and animal one has to quantify the partial use of the component parts. Total availability (G) is defined as the product of absorbability (A) and metabolic efficiency (V). When the intake of a trace element is represented by the quantity b the quantity v which is finally usable by the body for essential metabolic functions can be calculated by the following equation:

$v = b \cdot A \cdot V$ or $v = b \cdot G$ (Kirchgessner et al., 1974).

Since most essential elements are involved in a great many metabolic functions and different factors x_i and z_j will influence the absorbability and metabolic efficiency, it is difficult to determine the quantity v experimentally. In the report by Kirchgessner et al. (cit. above, 1974) some model studies are presented which are outlined also in this paper but in connection with other discussions.

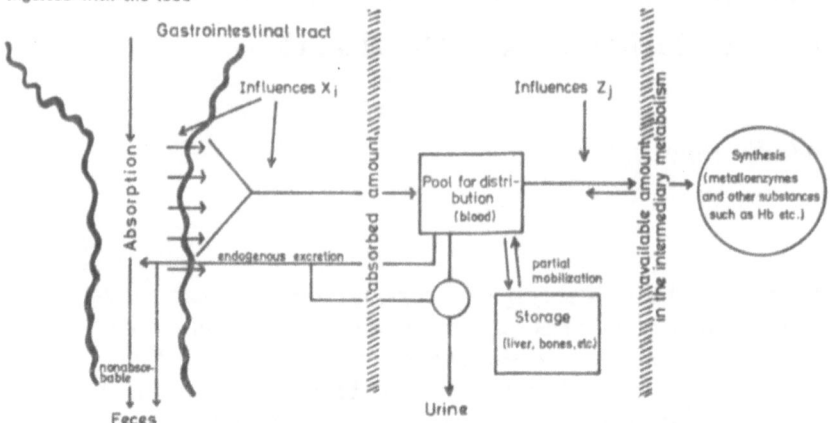

Essential trace elements
ingested with the food

Fig. 1. Scheme for partitioning the utilization of trace elements into component parts.

1. ESSENTIAL FUNCTIONS – PHARMACOLOGICAL ACTION

In this symposium much is being said about the toxicity of elements. Thus, before the concept of Fig. 1 is discussed in more detail, a few aspects shall be presented first to mark off the action of trace elements ranging from important roles as essential factors in metabolism to pharmacodynamic actions and further to toxicity. A trace element is considered essential, especially when its absence in the diet leads to deficiency symptoms, and a supplementation of this element can prevent or reverse these abnormalities. This is certainly very difficult to show because 'removing trace elements' from food requires rather involved analytical procedures. Basically, however, it is sufficient to prove that the element in question has an essential function in metabolism. The definition of an essential trace element given by Cotzias (1967) may be broader.

In Fig. 2 and 3 one could see clinical deficiency symptoms of zinc. Since most human and animal foods contain trace elements at certain levels, synthetic or semisynthetic diets are developed for appropriate experimental studies. Weaned rats were fed a semisynthetic casein diet with 1.2 ppm zinc. At the end of the depletion after 30 days the Zn-deficient rats have a weight of 48 g in comparison to pair-fed and ad lib. controls of 84 and 182 g respectively and show severe deficiency symptoms (Fig. 2, Kirchgessner and Roth, 1974b). When zinc is added to the

16

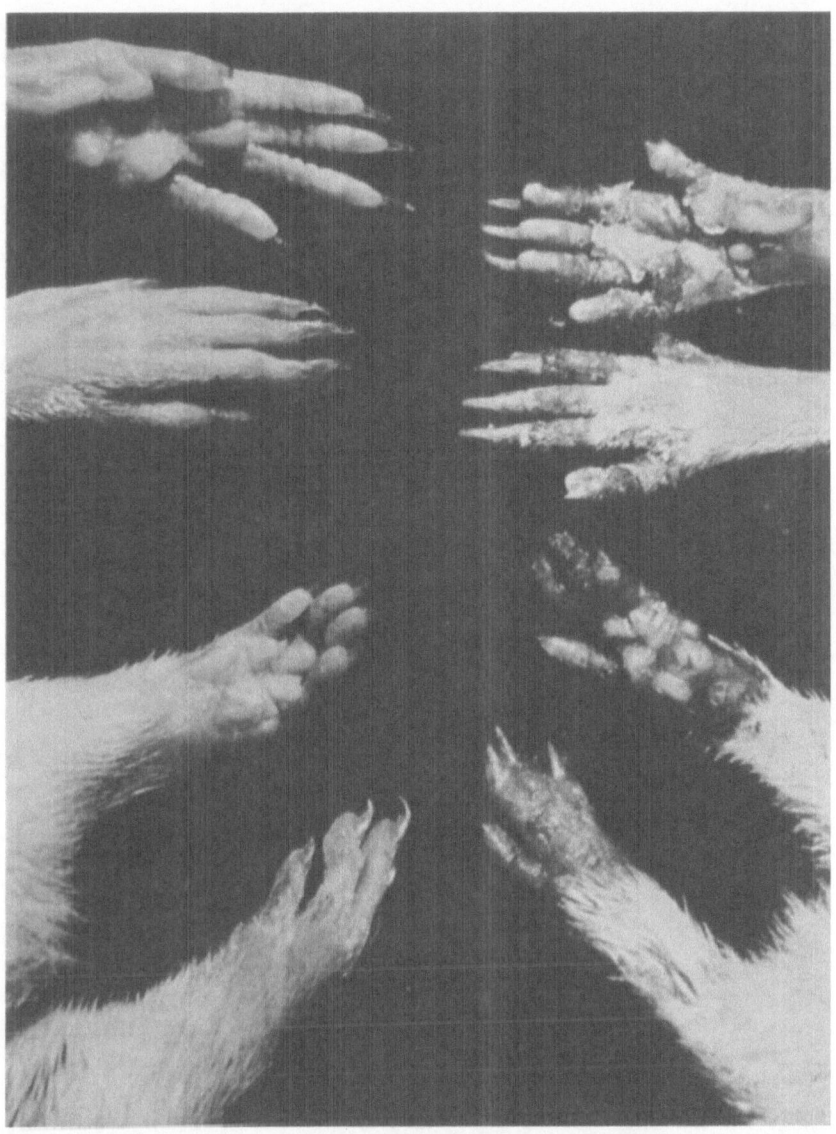

Fig. 2. Zn deficiency symptoms on extremities of growing rats in comparison to pair-fed controls (Zn deficiency below, controls above).

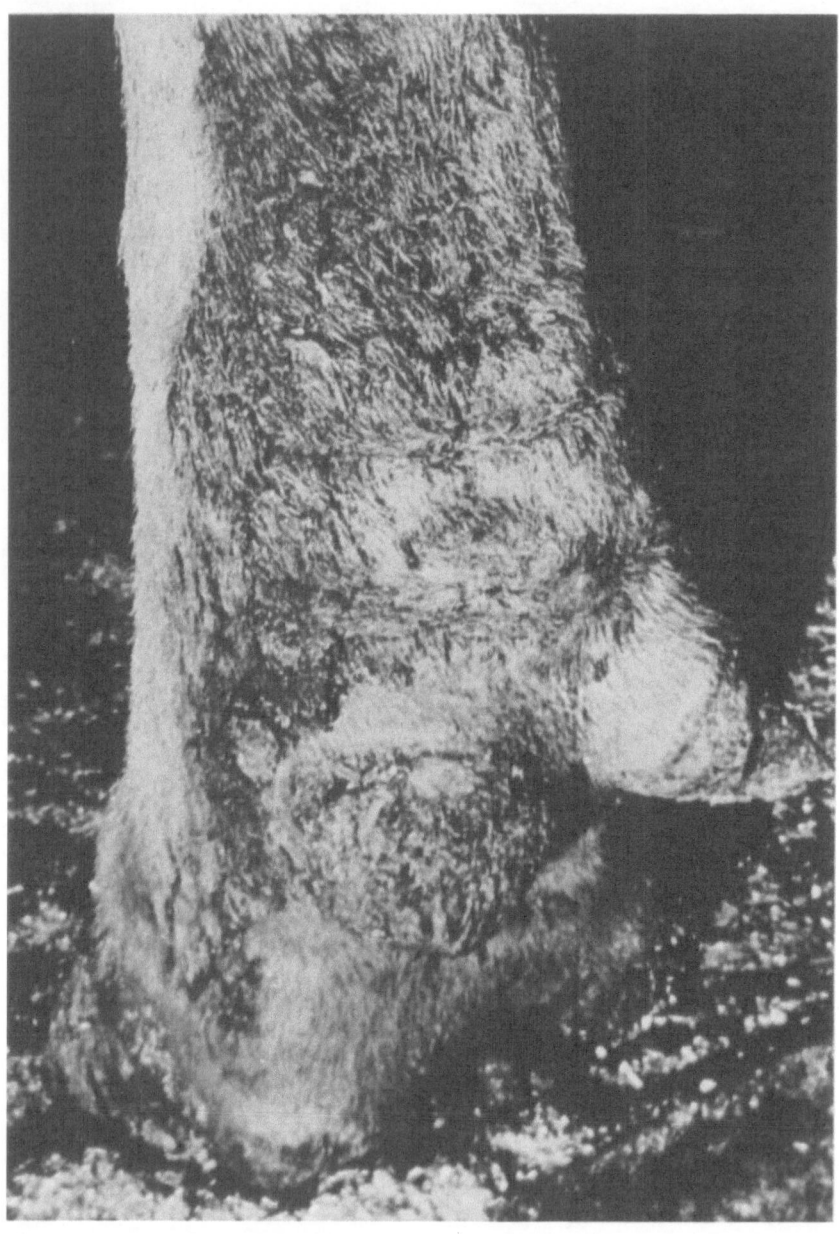

Fig. 3. Experimental parakeratose on the lower hind leg of a dairy cow in advanced state of Zn deficiency.

deficient diet, the animals recover within a few days and regain a healthy appearance within two weeks. Similar studies can be conducted also using farm animals. Dairy cows were fed a semisynthetic diet. The cows, once they are adapted to such rations, give birth to healthy calves and produce milk at optimal levels provided that no essential nutrient is missing in their diet. The lactating cows, however, which get the Zn-deficient diet, show parakeratotic Zn deficiency symptoms (Fig. 3, Schwarz and Kirchgessner, 1974).

Of course, we have to distinguish specifically between the importance of the trace element as an essential factor for the metabolism to function properly and its pharmacodynamic action. Supplying an essential trace element in excess of its requirement, therefore, cannot contribute to an increased performance and improved health. This will be pointed out with zinc as an example (Fig. 4). Weanling rats fed a semisynthetic,

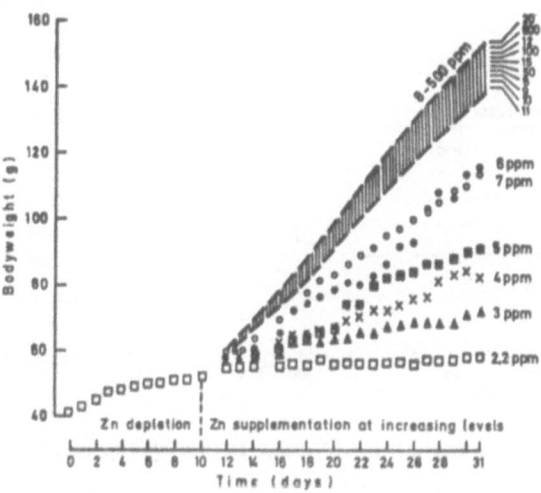

Fig. 4. Average body weight gain of weaned rats during Zn supplementation after a short period of Zn depletion.

partially purified casein diet (cf. Pallauf and Kirchgessner, 1971a) and Zn-depleted for a short time before Zn supplementation showed optimum weight gains with 8 ppm zinc, while higher doses were of no additional benefit (Pallauf and Kirchgessner, 1971b). After a supplementation for three weeks (Fig. 5) the serum had optimum zinc content with 12 ppm dietary zinc, and liver and bones with 15 ppm dietary zinc (Kirchgessner and Pallauf, 1972b; Pallauf and Kirchgessner, 1972a). A supplementation

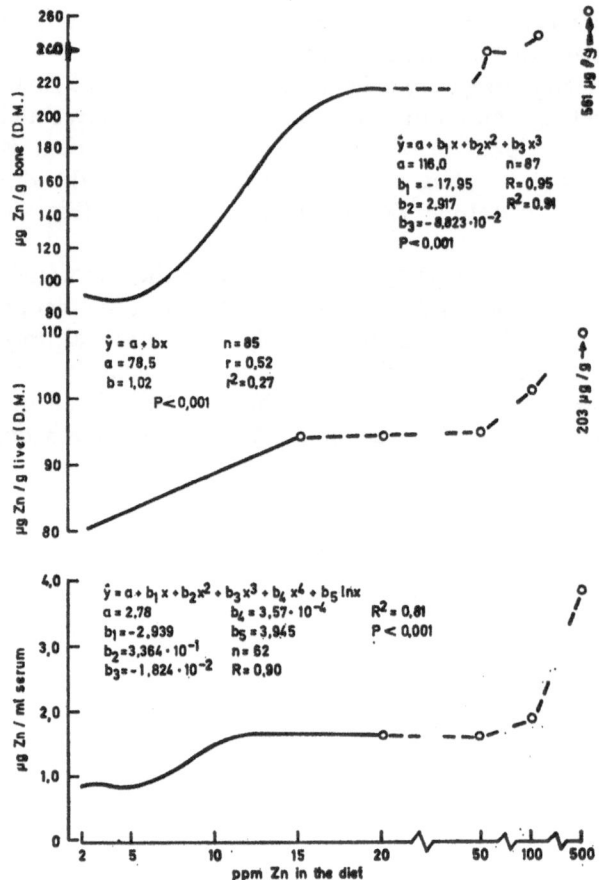

Fig. 5. Regression of Zn concentration in different tissues of repleted rats on the dietary Zn level.

exceeding these levels does not change the zinc status of the organs any more because of homeostatic regulation. Thus, according to these studies the requirement is fully met with 12 to 15 ppm zinc in the diet. Only extremely high supplements (500 ppm zinc) which exceed the regulatory capacity of the animal body lead to prepathological accumulation of zinc in the tissues. A pharmacological mode of action surely exists during very high Cu doses in swine production or in the effect of high amounts of lithium in manicdepressive states in human medicine (Mertz, 1971).

2. TRACE ELEMENT ABSORPTION

In meeting the optimum requirement for trace elements the content as such, however, is less crucial than the portion present that can be absorbed by man or animal. Therefore, the requirement should be used with reference to the absorbable amount as is done in the case of nutrients. While the digestibility of nutrients can be measured readily, it is quite different, however, with trace elements just as it is with minerals. In contrast to nutrients, elements after their absorption are returned in part to the feces with the bile (Mahoney et al., 1955, Gitlin et al., 1960, Scheuer and Barka, 1964), via the pancreas (Magee and Hong, 1959), and through the intestine (Mahoney et al., 1955, Owen, 1964; Bertinchamps et al., 1966, Methfessel and Spencer, 1973). Apart from the absolute level in the diet, a number of other factors influence the absorption of trace elements.

In the gastrointestinal tract the trace elements interact with the gastric, pancreatic, and biliary secretions and also with the other dietary constituents. Thereby, the pH, the solubility, and the capactity to form metal complexes or chelates play a special role. Then one or more of the coordination positions of the central atom is enclosed by ligands as if being held by crab pincers. The thermodynamic and kinetic stability of these complexes influence the absorption. Certainly, the metals from chelates of a lower stability are absorbed better. How important a role the pH can play in changing the absorption is demonstrated in our studies with ruminants (Kirchgessner, 1959) on the dependence of Cu absorption on the Ca content of the ration, an effect already noted by Tompsett (1940). As the Ca content of the ration increases, utilization of dietary copper by the ruminant diminishes sharply. Dietary calcium stems to a large extent from compounds with an alkaline reaction which subsequently lower the acidity of the intestinal contents and cause the pH to rise. In our in vitro experiments we found that dialysis of the Cu aquoion ($CuSO_4$) decreases sharply above pH 5.5 because copper is precipitated as hydroxide (Weser and Kirchgessner, 1965b). This situation does not apply to pigs and chickens because the pH of their stomachs is considerably lower (Kirchgessner et al., 1960b).

Dietary phosphates also may greatly influence the absorption of trace elements. Thus, the absorption of iron (Büttner and Muhler, 1959; Forth et al., 1965; Forth and Rummel, 1966), of zinc (Lewis et al., 1957; Heth et al., 1966), and of copper (Kirchgessner and Weser, 1965) is affected adversely by the addition of various phosphates. Among the organic phosphates phytate exerts the strongest effect.

Generally, trace elements appear to be absorbed better during deficiency than during a normal supply. Forth et al., (1968) showed that dietary iron was utilized better by anaemic rats than by animals supplied adequately with iron. According to our in vitro and in situ investigations, when copper, zinc or iron were deficient, these elements were also better absorbed than in the controls (Fig. 6, Kirchgessner et al., 1973; Schwarz and Kirchgessner, 1973; Schwarz and Kirchgessner, 1974a, b, c). Additionally

Intestinal transfer	Dietary pretreatment		
	Cu depletion	Zn depletion	Fe depletion
Cu	↑	↑	± 0
Zn	± 0	↑	± 0
Fe	↓	± 0	↑

↑: increase P<0,01;   ↓: decrease P<0,01;   ±0: no significant differences

Fig. 6. Intestinal transfer of Cu, Zn or Fe after Cu, Zn or Fe depletion.

Fig. 6 shows an interaction between Zn depletion and an increased Cu absorption and Cu depletion and a lowered Fe absorption (Schwarz and Kirchgessner, 1974 c).

Trace element complexes

There is no doubt that the type of complex or chelate in which trace elements are present in the diet, or the breakdown products which they form in the gastrointestinal tract, play a major role in absorption. The great importance of the complexes in food was demonstrated by Mills et al. (1954, 1956, 1958). He showed that copper from a complex present in meadow grass can be absorbed considerably better than from $CuSO_4$. Similar complexes also exist in the case of zinc and manganese (Bremner, 1970).

How strongly dietary constituents can influence absorption became apparent through in vitro experiments (Weser and Kirchgessner, 1965 a, b, c). The dialysis of copper sulphate was inhibited by dietary constituents. An explanation for this observation is that the rate of diffusion is diminished by the formation of complexes. The inhibition increased with the extent of formation and the stability of the complexes. Corresponding

22

results were obtained under analogous conditions in vivo (Kirchgessner and Weser, 1965). Fasting animals absorbed the copper from copper sulphate considerably faster than from Cu complexes.

Just the opposite was true during food intake. The high concentration of Cu-binding substances in the intestinal tract generally leads to the formation of macromolecular compounds with the soluble copper of the food. Consequently, the rate of Cu transfer is considerably reduced. Under these conditions copper is absorbed better from small stable chelates because it cannot be bound by macromolecular ligands (Kirchgessner and Weser, 1965; Kirchgessner and Grassmann, 1970 a). Kratzer and Starcher (1963) and Nielsen et al. (1966) were able to show that zinc could be utilized from its complexes which have a thermodynamic stability constant of up to $\log K = 18$. At least, the Zn deficiency symptom could be reversed with the administration of such Zn complexes.

Amino acid complexes

In contrast to the minerals Na, K, and to about half also Ca and Mg, trace elements are bound almost completely to organic ligands. Certainly, in biological media protein and its products of hydrolysis, the amino acids, occur most frequently as ligands. With this in mind we devoted a series of experiments (Kirchgessner, et al., 1967; Kirchgessner and Grassmann, 1970 a, b) to investigate the effect of amino acids and some 'derivatives' on Cu absorption. When copper was added as an amino acid, peptide, or polypeptide complex, the Cu content of the liver was considerably higher than in the case of copper sulphate. The Cu complexes of monomeric amino acids are absorbed better than those of dimeric ones. The latter ones, in turn, are absorbed better than those of trimeric or polymeric amino acids. Besides stability and size, the configuration of the amino acid affects the rate of absorption. When Cu-D-amino acid complexes were ingested, considerably less copper was stored in the liver than when the corresponding L-compounds were administered. Still another factor of influence is the type of amino acid. Thus, copper from complexes of the leucine series, for example, was absorbed better throughout than from the much smaller alanine complexes.

These results also indicate that the influence of molucular size on absorption is not always strictly valid. Lengthening the alkyl residue of an amino acid, therefore, cannot be taken as the sole criterion of the rate of absorption of the corresponding Cu complex. These indications of a specific influence of the amino acid on Cu absorption were investigated in the case of 15 different L-amino acid complexes (Kirchgessner and Grassmann,

1970 a). Again, the Cu supplementation resulted in a sharp rise in the Cu content in the liver of all groups. When the Cu storage in the liver of the group receiving copper sulphate is set equal to 100, the retention in the other groups is on the order of 80 to 140. It is particularly high when Cu-(L-Val)$_2$, Cu-(L-Phe)$_2$, or Cu-(L-Ile)$_2$, and also Cu-(L-Tyr)$_2$ or Cu-(L-Leu)$_2$ are supplemented. The storage in the liver is the greatest with supplements of complexes of essential amino acids. Thus, within the range of the stability constants of the amino acid complexes examined, the specific effect of the ligands is no doubt of primary importance. Similarly, it can be deduced from the findings that the molecular size, within certain limits, is less important than the specific influence of the ligand (Kirchgessner and Grassmann, 1970 a). In spite of the very different size of the ligands, the Cu retention from phenylalanine (mol wt 165) and valine (mol wt 117) was more or less of the same order but considerably higher than from alanine (mol wt 89) and threonine (mol wt 119). These in vivo observations could be confirmed in vitro by the intestinal Cu transfer from the mucosal to the serosal solution and by the intestinal Cu uptake of the intestinal wall for several Cu amino acid complexes (Grassmann et al., 1971, Schwarz et al., 1973). Improved Cu absorption was also found during supplementation of Cu complexes with organic acids (Grassmann and Kirchgessner, 1969).

3. METABOLIC EFFICIENCY OF TRACE ELEMENTS IN METABOLISM

Not merely the absorption but also the metabolic efficiency of the trace elements depends upon the formation of complexes. Here, of course, the stability of the chelates is of major importance. The prerequisite for a trace element to go into function is its release from the complex and its exchangeability, respectively. Certainly, this can be measured only by biochemical criteria. The metal-containing enzymes present themselves as particularly suitable tests. For measuring differences in the metabolic efficiency of copper ceruloplasmin is suited. In Tables I and II the changes of the Cu content and the ceruloplasmin activities of the liver are shown in response to Cu depletion and supplementation with various Cu complexes (Kirchgessner and Grassmann, 1970 c). With decreasing liver Cu storage the ceruloplasmin activity was reduced concurrently (Table I). After the depletion the food was supplemented with various Cu complexes (Table II). Judging from the ceruloplasmin activity, it seems that in this experiment copper from Cu-L-leucinate has the best metabolic efficiency. Copper from Cu-fumarate has the lowest metabolic efficiency. The

Table I. Cu content of the liver and ceruloplasmin activity during Cu-deficient nutrition for 40 days

Time (days)	Cu content of the liver ($\mu g/g$ D.M.)	Ceruloplasmin activity ($E_{550}/10$ min/ml Serum)
0	30.7 ± 8.6	0.332
10	15.4 ± 1.3	0.228
17	10.5 ± 1.0	0.232
26	10.9 ± 1.0	0.140
40	6.8 ± 0.8	0.104

Table II. Effect of different Cu compounds on the Cu content of the liver and the ceruloplasmin activity

Supplement	Cu content (μg/total liver)	Ceruloplasmin activity ($E_{550}/10$ min/ml Serum)
$CuSO_4$	34.9 ± 5.5	0.189 ± 0.057
Cu-citrate	34.7 ± 5.5	0.240 ± 0.044
Cu-fumarate	37.8 ± 1.8	0.143 ± 0.044
Cu-oxalate	39.1 ± 4.2	0.257 ± 0.050
Cu-EDTA	42.2 ± 6.1	0.313 ± 0.065
Cu-L-leucinate	38.0 ± 2.7	0.310 ± 0.050

Table III. Activities of alkaline phosphatase in serum of depleted, repleted and control rats

days	control rats (96 ppm Zn)	pair-fed rats (96 ppm Zn)	deficient rats (1.2 ppm Zn)	repleted rats (4.5 ppm Zn)	repleted rats (12 ppm Zn)
		activity of alkaline phosphatase (mU/ml serum)			
0	–	88 ± 11	–	–	–
14	217 ± 23	246 ± 32	55 ± 3	–	–
17/3	222 ± 23	255 ± 42	51 ± 7	77 ± 14	209 ± 46
22/8	182 ± 37	238 ± 29	55 ± 17	83 ± 21	218 ± 45
29/15	201 ± 27	215 ± 16	50 ± 7	96 ± 17	221 ± 38

influence of the other ligands showed the same trend in the liver storage test.

Besides ceruloplasmin, the cytochrome oxidase also is reduced in the liver of pigs and rats fed Cu-deficient diets (Gubler et al., 1957; Dreosti, 1967). Similarly, the activity of amine oxidase (Hill and Kim, 1967) and tyrosinase (Frieden, 1968) may depend on copper nutrition.

Also the activity of various Zn-depended enzymes is altered during zinc deficiency. In serum the activity of the alkaline phosphatase of depleted rats was lowered by 27 % after 2 days compared with ad libitum fed control animals, by 48 % after 4 days and by 75 % at the end of the 30-days experiment (Roth and Kirchgessner, 1974a). On the other hand the lactate and malate dehydrogenase in serum, did not show any changes in activity throughout this study. In a second experiment (see Table III) the activity of the alkaline phosphatase, in succession to a 14-day depletion (1.2 ppm Zn) increased only slightly after feeding a diet with 4.5 ppm Zn, but reached the value of pair-fed control animals (96 ppm) within 3 days after feeding a diet with 12 ppm Zn. The activity of the alkaline phosphatase in rat femur showed a similar response to Zn depletion and repletion in comparison to that of the serum study, except that the activity increased and decreased somewhat slower (Roth and Kirchgessner, 1974b). The activities of malate dehydrogenase in liver (Roth and Kirchgessner, 1974c) and of lactate and alcohol dehydrogenase in muscle (Roth and Kirchgessner 1974 d) remained unaltered during zinc depletion. The liver alcohol dehydrogenase (Roth and Kirchgessner 1974 c), however, was depressed by 26% and the muscle malate dehydrogenase (Roth and Kirchgessner, 1974 d) by 24% in the state of severe zinc deficiency. A significantly lower activity of the liver lactate dehydrogenase, – amounting to 34 % of the activity of ad libitum fed controls after 10 days of the experiment and to 58 % after 30 days – was also observed in both pair-fed control and Zn-depleted groups (Roth and Kirchgessner, 1974 e). Therefore, this was not a consequence of zinc deficiency per se. The pancreas lost 24 % of its carboxypeptidase A activity after 2 days of Zn depletion, 37 % after 10 days and 47 % at the end of the 29 days experiment (Roth and Kirchgessner, 1974 e). Dietary repletion at a level of 4.5 ppm Zn did not increase the activity of the pancreas carboxypeptidase (see Table IV), whereas after three days of feeding a diet with 12 ppm Zn an activity was reached which was comparable to that of pair-fed animals (96 ppm Zn). Furthermore, after 29 days of depletion in this experiment it was possible to demonstrate for the first time a reduction in the activity of the pancreas carboxypeptidase B compared with ad libitum (by 52 %) and pair-fed (48 %) control animals.

Based on our observations, only the alkaline phosphatase in serum and

Table IV. Activities of pancreatic carboxypeptidase A of depleted, repleted and control rats

| days | activity of pancreatic carboxypeptidase A (OD/min/ mg protein) | | | | |
	control rats (96 ppm Zn)	pair-fed rats (96 ppm Zn)	deficient rats (1.2 ppm Zn)	repleted rats (4.5 ppm Zn)	repleted rats (12 ppm Zn)
0	–	0.079 ± 0.015	–	–	–
14	0.058 ± 0.005	0.053 ± 0.008	0.030 ± 0.005	–	–
17/3	0.051 ± 0.010	0.045 ± 0.006	0.030 ± 0.006	0.027 ± 0.004	0.042 ± 0.008
22/8	0.061 ± 0.014	0.062 ± 0.014	0.031 ± 0.003	0.028 ± 0.005	0.053 ± 0.006
29/15	0.051 ± 0.004	0.066 ± 0.009	0.023 ± 0.002	0.028 ± 0.004	0.052 ± 0.005

femur as well as the carboxypeptidase A and presumably also B in pancreas appear commendable for measuring the metabolic efficiency of zinc in future studies. This results on the activities of zinc metalloenzymes in response to depletion and repletion of zinc are partly in agreement with the studies of Hsu et al. (1967), Oberleas et al. (1969), Prasad et al. (1971) and Mills et al. (1967), but, on the other hand are also contrary to observations made by Luecke et al. (1968) and Hsu et al. (1966).

4. DETERMINATION OF DEFICIENT OR SUBOPTIMAL SUPPLY OF TRACE ELEMENTS

These enzyme activities, however, can be used not only as a mean to measure the metabolic efficiency of vastly different trace element compounds and requirements of trace elements (Kirchgessner and Roth 1974a), but also as an excellent indicator of the specific state of supply of trace elements. Thus, cytochrome oxidase (Mills and Dalgarno, 1970; Poole, 1970) and ceruloplasmin (Todd, 1970; Rish, 1970), already have been proposed as tests for copper. For zinc there will be alkaline phospoatase in serum and bones, sometimes also carboxypeptidase A and B in the pancreas as a possibility for determination of suboptimal supply (Kirchgessner and Roth, 1974a). Certainly, such determinations together with serum analyses and biopsies of liver and bones are the best way to evaluate the supply state of the animal body. Fig. 7 shows the change of

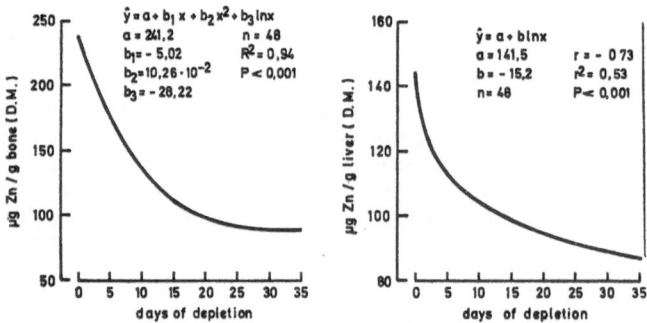

Fig. 7. Zn depletion in livers and bones (femore and humeri) of weaned rats under extreme dietary Zn deficiency (2 mg Zn/kg diet).

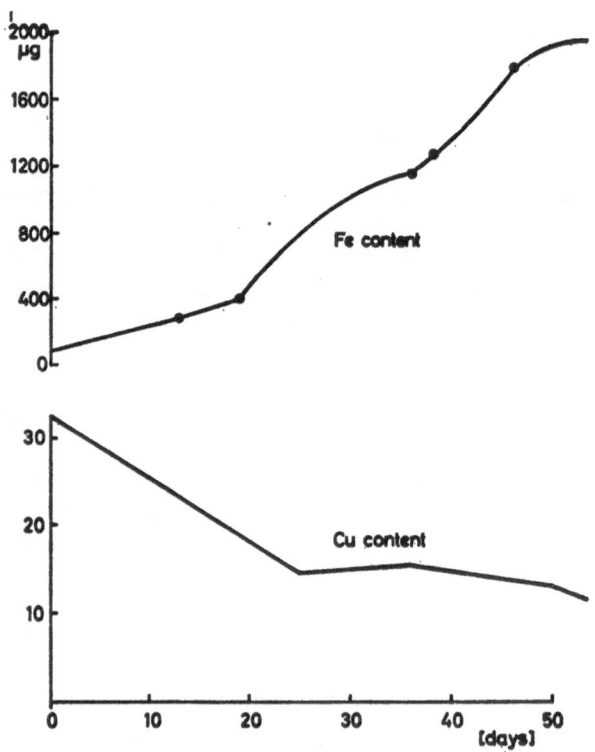

Fig. 8. Response of the Cu and Fe content of the total liver to deficient supply.

28

the zinc contents of liver and bones (femura and humeri) of growing rats during extreme Zn depletion (Kirchgessner and Pallauf, 1972a).

5. TRACE ELEMENT INTERACTIONS

The metalloenzymes can be used not only for measuring the metabolic efficiency, the actual supply state, the requirement and the essentiality of trace elements, but also may explain interactions observed between trace elements. The interaction observed between iron and copper can demonstrate this. In the liver of the rat iron accumulates very rapidly in response to a deficient supply of copper (Fig. 8) and the plasma iron content decreases. This is true even during a 'mild' Cu deficiency (Grassmann and Kirchgessner 1973 a, b). This disturbance of the Fe utilization, which in recent studies was attributed to an impeded mobilization of iron due to a ceruloplasmin deficiency (Frieden, 1970), slowly results in a small but yet significant decline in the hemoglobin level and catalase activity of the blood. When ceruloplasmin is injected, the plasma iron is increasing within a few hours already, while the iron concentration of the liver decreases (Grassmann and Kirchgessner, 1973a). The oxidase activity of ceruloplasmin seems to be responsible for the moblilization of the stored iron.

The deficiency symptoms of B-vitamins and some trace elements look similar to those of Zn deficiency. Thus, we studied whether these vitamins and the trace elements Mn, Cu, Fe, Co and Ni can possibly exert replacement functions of Zn. The vitamins biotin, folic acid, B_1, B_2, B_6, B_{12}, pantothenic acid and niacin did not alleviate the symptoms of zinc deficiency nor did they affect the zinc depletion in serum and liver (Pallauf and Kirchgessner, 1972c, 1973). Similarly, the additions of the trace elements mentioned had, under zinc deficiency, no effect on symptoms and live weight (Kirchgessner and Pallauf, 1973; Pallauf and Kirchgessner, 1974a). The influences on the zinc status are shown in Fig. 9 (Pallauf and Kirchgessner, 1974b). There evidently are interactions between zinc and some of the trace elements in the intermediary metabolism. But there were no indications, however, for replacement functions of these trace elements added during zinc deficiency.

SUMMARY

The actual intake of trace elements with the food is less important than the absorbable amount in providing an adequate dietary supply of essential

criteria	experiment	Mn				Cu				Fe			Co			Ni		
		120	300	1500	3000	12	20	40	200	120	400	2000	3	10	50	3	10	50
liver: µg total Zn	1	0	++	++	+++	0	0	0	++	++	++	++	0	-	0	0	0	0
	2	++	+++	0	+++	0	0											
µg Zn/g	1	++	++	++	+++	0	0	0	0	0	++	+++	0	0	0	++	0	+++
	2	+++	0	0	+++	0	0											
serum: µg Zn/ml	1	0	0	+++	+++	0	0	---	0	0	0	0	0	0	0	--	--	--
	2	0	0	0	+++	0	0											

0 = not significantly different from the control
+ = significantly different (P≤0,05), increase <15 per cent ;
++ = "   "   "   , " 15-30per cent ;
+++ = "   "   "   , " >30per cent ;
− = significantly different (P≤0,05), decrease <15 per cent
−− = "   "   "   , " 15-30per cent
−−− = "   "   "   , " >30per cent

Fig. 9. Influence of various trace elements added at different levels to a low zinc diet on the zinc status of liver and serum.

trace elements in man and animal. The absorption, however, is strongly influenced, besides other factors, by the pH in the intestinal tract and various ligands in the chymus, and also by the supply of the organism with the respective trace element. The absorption of copper, for example, is significantly improved by the specific influence of different amino acids. The complex formation determines not only the absorption, but also the metabolic efficiency of the trace elements. The metabolic efficiency can be determined by the activity of metalloenzymes. These enzymes can also be used for the determination of the supply of the body with the trace elements and possible metabolic interactions of the elements. The time course of the Zn contents in liver and bones of growing rats after depletion and repletion is used as an example to show that biopsy can be used for additional studies on the supply of trace elements.

REFERENCES

Bertinchamps, A.J., S.T. Miller, and G.L. Cotzias, 'Interdependence of routes excreting manganese', *Am. J. Physiol.* **211** (1966) 217.
Bremner, I. in C.F. Mills (ed.), *Trace Element Metabolism in Animals*, (Edinburgh, Livingstone, 1970), p. 366.
Büttner, W. and J.C. Muhler, 'Effect of dietary iron on phosphate metabolism', *Proc. Soc. Exper. Biol. Med.* **100** (1959) 440.
Cotzias, G.C., 'Importance of trace substances in environmental health as exemplified by manganese', in D.D. Hemphill (ed.), *Trace Substances in Environmental Health-I*, (Columbia, University of Missouri, 1967) p. 5.

Dreosti, I.E., 'Cytochrome oxidase activity in copper deficient adult rats,' *S. African J. Agric. Sci.* **10** (1967) 95.

Forth, W., K. Pfleger, W. Rummel, E. Seifen and S.J. Richmond, 'Der Einfluss verschiedener Liganden auf Resorption, Verteilung und Ausscheidung von Eisen nach oraler Verabfolgung,' *Arch. exp. Path. Pharmak.* **252** (1965) 242.

Forth, W. and W. Rummel, 'Abhängigkeit der Eisenresorption von der Eisenbindung durch den Darm,' *Med. Pharmacol. exp.* **14** (1966) 384.

Forth, W., W. Rummel, and K. Pfleger, 'Der Einfluss von Liganden auf die Retention von Eisen nach oraler Verabfolgung an normale und anämische Ratten', *Arch. Exp. Path. Pharmak.* **261** (1968) 225.

Frieden, E., 'The biochemistry of copper', *Sci. Am.* **218** (1968) 103.

Frieden, E., 'Ceruplasmin, a link between copper and iron metabolism', *Nutr. Rev.* **28** (1970) 87.

Gitlin, D., W.J. Hughes, and C.A. Janeway, 'Absorption and excretion of copper in mice', *Nature* **188** (1960) 150.

Grassmann, E., and M. Kirchgessner, 'Kupfer-Absorption aus Komplexen mit verschiedenen organischen Säuren', *Z. Tierphysiol. Tierernährg. Futtermittelk.* **25** (1969) 125.

Grassmann, E. and M. Kirchgessner, 'Zur Eisenverwertung bei unterschiedlicher Kupferversorgung', *Archiv Tierernährung* **23** (1973 a) 261.

Grassmann, E. and M. Kirchgessner, 'Zur Fe- und Cu-Verfügbarkeit im Stoffwechsel bei unterschiedlicher Fe- und Cu-Versorgung', *Z. Tierphysiol. Tierernährg. Futtermittelk.* **31** (1973 b) 113.

Grassmann, E., A. Wetzstein and M. Kirchgessner, 'Untersuchungen am isolierten Rattendarm über Durchgang und Bindung von Kupfer', *Z. Tierphysiol. Tierernährg. Futtermittelk.* **28** (1971) 28.

Gubler, C.J., G.E. Cartwright and M.M. Wintrobe, 'Studies on copper metabolism. Enzyme activities and iron metabolism in copper and iron deficiency', *J. Biol. Chem.* **224** (1957) 533.

Heth, D.A., W.M. Becker and W.G. Hoekstra 'Effect of calcium, phosphorus and zinc on zinc-65 absorption and turnover in rats fed semipurified diets', *J. Nutr.* **88** (1966) 331.

Hill, C.H, and C.S. Kim, 'The derangement of elastin synthesis in pyridoxine deficiency', *Biochem. Biophys. Res. Commun.* **27** (1967) 94.

Hsu, J.M., J.K. Anilane and D.E. Scanlan, 'Pancreatic carboxypeptidases: activities in zinc deficient rats', *Science (N.Y.)* **153** (1966) 882.

Hsu, J.M. and J.K. Anilana, 'Effect of Zn deficiency on Zn metallo-enzymes in rats', *Proc. Seventh Int. Con. Nutr.* **5** (1967) 753.

Kirchgessner, M., 'Wechselbeziehung zwischen Spurenelementen in Futtermitteln und tierischen Substanzen sowie Abhängigkeitsverhältnisse zwischen einzelnen Spurenelementen bei der Retention', *Z. Tierphysiol. Tierernährg. Futtermittelk.* **14** (1959) 278.

Kirchgessner, M. and E. Grassmann, 'Absorption von Kupfer aus den Cu (II)-L-Aminosäure-Komplexen', *Z. Tierphysiol. Tierernährg. Futtermittelk.* **26** (1970 a) 3.

Kirchgessner, M. and E. Grassmann, 'The dynamics of copper absorption', in C.F. Mills (ed.), *Trace Element Metabolism in Animals,* (Edinburgh, Livingstone, 1970-b) p. 277.

Kirchgessner, M. and E. Grassmann, 'Untersuchungen zur Kupferverfügbarkeit mit Coeruloplasmin als Testenzym', *Z. Tierphysiol. Tierernährg. Futtermittelk.* **26** (1970 c) 340.

Kirchgessner, M., H.L. Müller, E. Weigand, E. Grassmann, F.J. Schwarz, J. Pallauf and H.-P. Roth, 'Zur Definition und Bestimmung der Absorbierbarkeit, intermediären

Verfügbarkeit und Gesamtverwertung von essentiellen Spurenelementen', *Z. Tierphysiol. Tierernährg. Futtermittelk.* **34** (1974) 3.

Kirchgessner, M., W. Munz and W. Oelschläger, 'Der Einfluss einer CuSO₄-Zulage auf die Retention von Mengen- und Spurenelementen bei wachsenden Schweinen', *Arch. Tierernährg.* **10** (1960b) 1.

Kirchgessner, M. and J. Pallauf, 'Zinkdepletion wachsender Ratten. Leber, Knochen, Schwanz und Ganzkörper', *Z. Tierphysiol. Tierernährg. Futtermittelk.* **29** (1972 a) 65.

Kirchgessner, M. and J. Pallauf, 'Zinkrepletion in Serum und Leber wachsender Ratten', *Z. Tierphysiol. Tierernährg. Futtermittelk.* **29** (1972 b) 77.

Kirchgessner, M. and J. Pallauf, 'Zum Einfluss von Fe-, Co- bzw. Ni-Zulagen bei Zinkmangel', *Z. Tierphysiol. Tierernährg. Futtermittelk.* **31** (1973) 268.

Kirchgessner, M. and H.-P. Roth, 'Bestimmung der Verfügbarkeit von Zink im Stoffwechsel sowie Ermittlung des Zinkbedarfs mittels Aktivitätsveränderungen von Zink-Metalloenzymen', *Arch. Tierernährg.* **25** (1975) in press.

Kirchgessner, M., and H.-P. Roth, 'Beziehung zwischen klinischen Mangelsymptomen und Enzymaktivitäten bei Zinkmangel', *Zbl. Vet. Med. A.* **22** (1975 b) 14.

Kirchgessner, M., F.J. Schwarz, and E. Grassmann, 'Intestinal absorption of copper and zinc after dietary depletion', *Bioinorg. Chem.* **2** (1973) 255.

Kirchgessner, M. and U. Weser, 'Komplexstabilität und Kupfer-Absorption', *Z. Tierphysiol. Tierernährg. Futtermittelk.* **20** (1965) 44.

Kirchgessner, M., U. Weser and H.L. Müller, 'Absorption einiger mono-, oligo- und polymerer Aminosäure-Cu-Komplexe', *Z. Tierphysiol. Tierernährg. Futtermittelk.* **22** (1967) 76.

Kratzer, F.H. and B. Starcher, 'Quantitative relation of EDTA to availability of zinc for turkey poults', *Proc. Soc. Exp. Biol. Med.* **113** (1963) 424.

Lewis, P.K., W.C. Hoekstra and R.H. Grummer, 'Restricted Calcium feeding versus zinc supplementation for the control of parakeratosis in swine', *J. Anim. Sci.* **16** (1957) 578.

Luecke, R.W., M.E. Olman and B.V. Baltzer, 'Zinc deficiency in the rat: effect on serum and intestinal alcaline phosphatase activities', *J. Nutr.* **94** (1968) 344.

Magee, D.F. and S.S. Hong, 'Daily output of pancreatic juice and some dietary factors which influence it', *Am. J. Physiol.* **197** (1959) 27.

Mahoney, J.P., F.A. Bush, C.J. Gubler, W.H. Moretz, G.E. Cartwright, 'Studies on copper metabolism. XV. The excretion of copper by animals', *J. Lab. Clin. Med.* **46** (1955) 702.

Methfessel, A.M. and M. Spencer, 'Zinc metabolism in the rat. II. Secretion of zinc into intestine', *J. Appl. Physiol.* **34** (1973) 63.

Mertz, W., 'Spurenelementen. Ihre Bedeutung in der Ernährungswissenschaft und Medizin', *Fortschr. Med.* **89** (1971) 1191.

Mills, C.F., 'Copper complexes in grassland herbage', *Biochem. J.* **57** (1954) 603.

Mills, C.F., 'The dietary availability of copper in the form of naturally occurring organic complexes', *Biochem. J.* **63** (1956) 190.

Mills, C.F. and A.C. Dalgarno, 'An evaluation of tissue cytochrome oxidase activity as an indicator of copper status', in C.F. Mills (ed.), *Trace Element Metabolism in Animals*, (Edinburg, Livingstone, 1970) p. 465.

Mills, C.F., J. Quarterman and R.B. Williams, 'The effect of zinc deficiency on pancreatic carboxypeptidase activity and protein digestion and absorption in the rat', *J. Biochem.* **102** (1967) 712.

Nielsen, F.H., M.L. Sunde and W.C. Hoekstra, 'Effect of some dietary synthetic and natural chelating agents on the zinc-deficiency syndrome in the chick', *J. Nutr.* **89**

(1966) 35.

Oberleas, D. and A.S. Prasad, 'Growth as effected by zinc and protein nutrition', *Am. J. Clin. Nutr.* **22** (1969) 1304.

Owen C.A., 'Absorption and excretion of ^{64}Cu–labeled copper by the rat', *Am. J. Physiol.* **207** (1964) 1203.

Pallauf, J. and M. Kirchgessner, 'Experimenteller Zinkmangel bei wachsenden Ratten', *Z. Tierphysiol. Tierernährg. Futtermittelk.* **28** (1971 a) 128.

Pallauf, J. and M. Kirchgessner, 'Zum Zinkbedarf wachsender Ratten', *J. Vit. Nutr. Res.* **41** (1971 b) 543.

Pallauf, J. and M. Kirchgessner, 'Zinkgehalte in Knochen und Ganzkörper wachsender Ratten bei unterschiedlicher Zinkversorgung', *Z. Tierphysiol. Tierernährg. Futtermittelk.* **30** (1972 a) 193.

Pallauf, J. and M. Kirchgessner, 'Zinkkonzentration in Blut und Serum wachsender Ratten bei Zinkmangel', *Zbl. Vet. Med. A* **19** (1972) 594.

Pallauf, J. and M. Kirchgessner, 'Zur Wirksamkeit von Biotin- und Folsäurezulagen bei Zinkmangel', *Int. J. Vit. Nutr. Res.* **42** (1972c) 555.

Pallauf, J. and M. Kirchgessner, 'Zur Wirksamkeit erhöhter Zulagen an Vitamin D_1, D_2, D_6, B_{12}, Pantothen- und Nikotinsäure bei Zinkmangel', *Int. J. Vit. Nutr. Res.* **43** (1973) 339.

Pallauf, J. and M. Kirchgessner, 'Effekt verschiedener Mn–, bzw. Cu–Zulagen bei mangelnder Zinkversorgung', *Zbl. Vet. Med. A* **21** (1974 a) 562.

Pallauf, J. and M. Kirchgessner, 'Zinc status in depletion and repletion and its relation to vitamins and trace elements', in W.G. Hoekstra, J.W. Suttie, H.E. Ganther and W. Mertz (eds.), *Trace Elements in Animal Metabolism –2*, (Baltimore, University Park Press, 1974-b), p. 534.

Poole, D.B.R., 'Cytochrome oxidase in induced hypocuprosis', in C.F. Mills (ed.), *Trace Element Metabolism in Animals*, (Edinburgh, Livingstone, 1970) p. 465.

Roth, H.-P. and M. Kirchgessner, 'Aktivitätsveränderungen verschiedener Dehydrogenasen und der alkalische Phosphatase bei Zink-Depletion und -Repletion', *Z. Tierphysiol. Tierernährg. Futtermittelk.* **32** (1974 a) 289.

Roth, H.-P. and M. Kirchgessner, 'Zum Einfluss unterschiedlicher Diätzinkgehalte auf die Aktivität der alkalischen Phosphatase im Knochen', *Z. Tierphysiol. Tierernährg. Futtermittelk.* **33** (1974 b) 57.

Roth, H.-P. and M. Kirchgessner, 'Zum Aktivitätsverlauf verschiedener Dehydrogenasen in der Rattenleber bei unterschiedlichter Zinkversorgung', *Z. Tierphysiol. Tierernährg. Futtermittelk.* **33** (1974 c) 1.

Roth, H.-P. and M. Kirchgessner, 'Zur Enzymaktivität von Dehydrogenasen im Rattenmuskel bei Zinkelmangel', *Z. Tierphysiol. Tierernährg. Futtermittelk.* **33** (1974 d) 67.

Roth, H.-P. and M. Kirchgessner, 'Zur Aktivität der Pankreas-Carboxypeptidase A und B bei Zink-Depletion und -Repletion', *Z. Tierphysiol. Tierernährg. Futtermittelk.* **33** (1974 e) 62.

Scheuer, P.J. and T. Barka, 'Effect of copper loading on uptake and excretion of copper-64 by rat liver', *Nature* **201** (1964) 1135.

Schwarz, F.J., E. Grassmann and M. Kirchgessner, 'Zur Absorption in vitro aus Cu-Aminosäure- und –Peptide – Komplexen', *Z. Tierphysiol. Tierernährg.Futtermittelk.* **31** (1973)98.

Schwarz, F.J. and M. Kirchgessner, 'Intestinale Cu–Absorption in vitro nach Fe- und Zu– Depletion. 11', *Z. Tierphysiol. Tierernährg. Futtermittelk.* **31** (1973) 91.

Schwarz, F.J. and M. Kirchgessner, 'Wechselwirkung bei der intestinalen Absorption von ^{64}Cu, ^{65}Zn und ^{59}Fe nach Cu–, Zn– oder Fe– Depletion', *Int. J. Vit. Nutr. Res.*

44 (1974 a) 116.

Schwarz, F.J. and M. Kirchgessner, 'Absorption von Zink-65 und Kupfer-64 im Zinkmangel', *Int. J. Vit. Nutr. Res.* **44** (1974 b) 258.

Schwarz, F.J. and M. Kirchgessner, 'Intestinal absorption of copper, zinc, and iron after dietary depletion', in W.G. Hoekstra, J.W. Suttie, H.E. Ganther and W. Mertz (eds.), *Trace Element Metabolism in Animals-2*, (University Park Press, Baltimore 1974 -b) p. 519.

Schwarz, W.A. and M. Kirchgessner, 'Experimenteller Zinkmangel bei laktierenden Milchkühen', *Vet. Med. Nachr.* in press (1975).

Todd, J.R. 'A survey of the copper status of cattle using copper oxidase (ceruloplasmin) activity of blood serum', in C.F. Mills (ed.), *Trace Element Metabolism in Animals*, (Edinburgh, Livingstone, 1970) p. 448.

Tompsett, S.L., 'Factors influencing the absorption of iron and copper from the alimentary tract', *Biochem. J.* **34** (1940) 961.

Weser, U. and M. Kirchgessner, 'Versuchsmethode zur Bestimmung der Dialysegeschwindigkeit von Kupferionen', *Z. Tierphysiol. Tierernährg. Futtermittelk.* **20** 1965 a) 34.

Weser, U. and M. Kirchgessner, 'Dialysegeschwindigkeit freier Kupfer-Ionen bei Zusatz verschiedener anorganischer Liganden', *Z. Tierphysiol. Tierernährg. Futtermittelk.* **20** (1965 b) 37.

Weser, U. and M. Kirchgessner, 'Einfluss organischer Liganden auf die Dialysegeschwindigkeit freier Kupferionen', *Z. Tierphysiol. Tierernährg. Futtermittelk.* **20** (1965c) 41.

Dr. C. J. A. Van den Hamer, Delft

Could you detect (e.g., immunologically) in copper-deficient animals – whose cerulplasmin level was lowered – any ceruloplasmin with less than its full complement of Cu (see, e.g., N. A. Holtzman and B. M. Gaumnitz. *J. Biol. Chem.* 245 (1970) 2350; C. J. A. Van den Hamer, this symposium, 1975).

Prof. Kirchgessner

So far we have no experiments assessing possible changes in the copper content of ceruloplasmin in response to copper deficiency. The results presented are based only on ceruloplasmin activity which, as was shown in many studies, is closely correlated to the ceruloplasmin concentration. We currently study the relationship between ceruloplasmin activity and the distribution of copper in blood plasma in response to the Cu status of rats. The results will soon be available.

Prof. I.G. Scheinberg, New York

Are the various Cu supplementation compounds – such as copper sulfate, copper oxalate, etc. – absorbed from the intestinal lumen as the compound (e.g., as Cu-oxalate) or do these compounds dissociate into cupric ions and the specific anions, which are then absorbed separately and not necessarily in equal molar amounts?

Prof. Kirchgessner

So far, we conducted no experiments to recover chelates in plasma. From our experiments with copper chelates of the natural amino acids, however, you may note that the ligand exerts a specific effect which cannot be explained on the basis of the stability of the copper chelates. Thus we can assume that copper must stand in relation to the ligands at the site of absorption or that the ligands must play a role in the absorption itself.

Unpublished data from our laboratory show that the spectral absorption maxima of a mixture of Cu^{++} and some amino acids shifted to shorter wave lengths with increasing pH. This indicates an interaction between copper (II) ions and amino acids above pH 3-4. In this connection, I like to call your attention to data from Mills who found that the pH of

35

the stomach contents of rats were maintained between 3.3 and 4.1 up to 17 hours after feeding.

Dr. H.G. van Eyk, Rotterdam

1. What is the mechanism regulating the Cu-Fe metabolism?
2. Are the small Cu-chelating complexes present as such in the body?
3. Are the stability constants of Cu-fumarate, -oxalate, etc. known?

Prof. Kirchgessner

1. At this time nothing is known about the regulation of the copper-iron interaction in metabolism. The role of copper seems to be related to its function in the two ferroxidases; the more important one of them is ceruloplasmin. From the influence of hormones on the activity of ceruloplasmin we can assume that in some way a hormonal regulation must be involved.

2. So far we conducted no experiments to recover the Cu chelates in vivo in the body. It is known that Cu amino acid complexes are present in the non-ceruloplasmin-bound Cu fraction of the plasma. But we do not know whether these chelates are synthesized in the blood as we assume to be the case for the Cu albumin fraction or whether this fraction results from the absorption of Cu amino acid chelates present or synthesized in the lumen (see also Scheinberg).

3. Regarding the stability constants we must distinguish between the thermodynamic and kinetic stability of the copper chelates. They together give the effective stability of the complexes. The kinetic stability depends upon the concentration of the competing ligands and trace elements in the intestinal lumen, or in the body-organs. Therefore we conducted our experiments under strictly constant conditions in order to eliminate this 'kinetic' influence as far as possible. Thermodynamic stability constants are of less interest. Values for most of the copper amino acid chelates are between 13 and 16, for Cu–EDTA it is about 18; for Cu–oxalate and –fumarate we found no values in the literature.

Physiology of toxic elements

J. Pařízek

Several factors have stimulated the rapidly growing interest in research on the toxic effects of certain trace elements. One of them is certainly the fact that the problems resulting from an inadequate uptake of trace elements and effecting human health or animal production have surpassed limits given by local, occupational or geographical conditions. It is important to note that under these conditions the changes in the exposure to certain trace elements are not necessarily confined to predilected groups of a given population. With such changes in exposure the possibility is substantially increased that those individuals could be affected who are particularly vulnerable namely (a) due to certain biological characteristics (connected e.g. with age or sex) or (b) due to the fact that the sensitivity has been modified by simultaneous or previous exposure to certain other environmental factors.

Another stimulus for research on the adverse biological effects of certain trace elements is connected with the emerging recognition of the possibility that changes in the exposure to certain trace elements could be of importance in relation to changes in the incidence of certain diseases (cf. Masironi 1972).

It is evident, and it should be mentioned particularly here, that progress of biomedical research in this area is, of course, dependent on the availability of reliable methods of chemical analysis, which would give sufficient information (sufficient in terms of accuracy and amount of data obtained) on trace elements in question present in the organism and its environment. It should be underlined, of course, that trace elements do not usually act as elements in the body and that different chemical compounds of the same trace element can have very different biological effects. This, together with the recognized significant transformation of compounds of certain trace elements in the organism and in the environment, should explain, why further development of analytical methods giving not only quantitative but also qualitative information should be considered as highly desirable.

However, I should like to turn the attention to another problem. The effects of trace elements within the organism are not dependent only on their chemical form, but also on the state of the living organism, as influenced, e.g. by the inherited characteristics including sex, by the stage of ontogenetic development of the organism, its previous history and by concomitant effects of other simultaneously acting factors of the environment.

The title of this paper was given as 'Physiology of toxic elements'. This title could sound, of course, controversial in several respects. As already mentioned, trace elements do not usually act as elements in the organism. Their biological effects are dependent not only on their chemical form but also on the dose: compounds of a certain trace element can be considered as essential or toxic, depending on the dose in question. However, even the same dose of the same compound of a certain trace element can be – under different conditions – connected with signs of toxicity or deficiency – depending on the physiological conditions of the organism, including nutritional status and supply of other trace elements (for further details and review cf. Pařízek 1972, 1975). It seems therefore that we should be more justified to speak of 'the role of physiology in studies on the toxicity of compounds of trace elements': it is, indeed, an important task for experimental medicine not only to discover new, so far unknown biological effects of important compounds of trace elements, but also to delineate those factors which play a decisive role influencing the character of the response of the organism, and the dose – effect relationship. It is evident that further progress in this field is essential for a better understanding of the biological significance of results obtained by analytical techniques, for the recognition of possibly so far overlooked environmental hazards and for the designation of those important additional factors, which should be followed up in epidemiological studies on possible associations between the incidence of certain diseases and the degree of exposure to a certain trace element.

The fact that a certain dose of a certain trace element (its compounds) can induce a specific lesion in a certain biological system (resulting e.g. in characteristic pathological changes localized selectively in a certain organ of the body) and the delineation of conditions on which the induction of this lesion depends, also represent useful sources of information on the biological system affected. Elucidation of factors influencing the toxicity of compounds of certain trace elements could help to understand the mechanisms of their distribution in the body, of their biotransformation and of their action in the system affected. As has been discussed previously elsewhere (Pařízek 1960, 1972), the cognitive value of studies on 'localized intoxications' in the sense envisioned by Claude Bernard in general terms more than a century ago is also fully valid in the field of trace elements.

In this sense, about 18 years ago, we started our studies on the toxicity of cadmium salts in relation to reproduction. Our first studies (Pařízek and Záhoř 1956, Pařízek 1957) revealed that parenteral injection of cadmium salts induced acute destruction of male gonads by haemorrhagic necrosis. This effect was quite consistent and specific for cadmium salts. Very recently Gabbiane et al. (1974) were able to detect dilatation of interendothelial clefts in the small vessels of the testes electronoptically as early as 15 min after injection of cadmium salts. In spite of all the work done so far in many laboratories in the world (for review see Friberg et al., 1974, Gunn and Gould 1970), the exact place and the biochemical mechanism of the primary action of cadmium in the testes and in certain other reproductive organs (see below) remains unexplained. However, during the years following our first reports on testicular haemorrhagic necrosis produced by cadmium salts (Pařízek and Záhoř 1956, Pařízek 1957) we have been able to show that cadmium has similar effects also in certain other organs connected with reproduction (Pařízek 1969). There is no time and space to go into details of all these experiments. However, I should like to use the case of the toxicity of cadmium in relation to reproduction to demonstrate how a toxic effect of a trace element can be dependent on the stage of development and function of a susceptible system and on certain other conditions of internal and external environment.

A dose of cadmium, fully effective in normal adult males (and also in hypofysectomized animals), also induced testicular destruction with resulting permanent infertility quite regularly when given to prepubertal rats at the age of two weeks after birth. However, these effects were not demonstrable when the same dose of cadmium was given to animals only few days younger, i.e. at the end of the first postnatal decade (Pařízek 1960). It was possible, therefore, to conclude that the effect of cadmium on the testes was not dependent on the presence of, or mediated through, the pituitary; on the other hand it was evident that the specific conditions responsible for the specific response to cadmium appeared in this organ at the end of the first decade of post-natal life.

Further experiments revealed that it was possible to induce an analogous sensitivity to cadmium in a non-responsive organ, namely in the ovary of adult rats. It is known, that administration of a dose of cadmium salts, resulting in permanent sterility of the male, does not affect the fertility of females in a similar way. In contrast to cadmium-induced haemorrhagic

39

necroses affecting the whole testis, similar damage of ovaries was not detected in females given an analogous parenteral injection of cadmium salts (Pařízek and Záhoř 1956). However, Kar et al. (1959) were able to induce severe ovarian lesions by cadmium salts given to prepubertal female rats. We have been able to confirm this observation and suspect that some phenomenon connected with ovulation could protect adult ovaries against cadmium. We decided therefore to test this hypothesis, using non-ovulating adult rats in which persistent oestrus had been induced by the injection of a high dose of androgens, given on the fifth day of postnatal life. In these rats cadmium salts given in adulthood induced profound haemorrhages and necroses in the female gonad, quite analogous to the effects of cadmium in the testes. Even here the effect was strictly specific for cadmium and was not induced by the injection of an equimolar amount of mercuric salts. Cadmium induced these changes quite consistently, and also in hypofysectomized rats. However, pre-treatment with gonadotrophic hormone (PMSG) completely prevented this effect (Pařízek et al., 1968a).

Further experiments revealed that cadmium salts given during pregnancy induced quite specific toxic effects, which cannot be demonstrated in non-pregnant animals. This could serve as an example of how an effect of a trace element can be strictly dependent on the functional state of the organism.

Parenteral injection of cadmium salts to pregnants rats (in a dose analogous to that inducing gonadal damage) results in rapid destruction of the foetal part of the placenta and in death of foetuses (Pařízek 1964). This quite consistent phenomenon is not connected with lethal effects affecting the maternal organism when cadmium in the dose mentioned is given to rats before the 17th day of pregnancy. However, when the same dose of cadmium salts is given during the later stage of pregnancy, the quite regular toxic effects in the foeto-placental unit, demonstrable in all cadmium treated pregnant rats (i.e., destruction of the foetal part of placenta and death of foetuses) are accompanied by lethal effects also affecting – in a high proportion of cases – the maternal organism. The specific lethal syndrome (induced in approximately two thirds of rats given cadmium during the last few days of pregnancy) can be characterized as a generalized Shwartzman reaction (Pařízek 1965).

It should be underlined that during this last period of pregnancy not only high lethality results from an exposure to cadmium in a dose which is survived well by non-pregnant females or males, but that the effects induced here are quite dissimilar to the effects produced by cadmium salts when given out of this period of peculiar sensitivity (Pařízek 1965). The severe renal damage, bilateral haemorrhagic renal necrosis, which is a characteristic part of this syndrome, is quite dissimilar from the renal effects of cadmium in non-pregnant animals. On the other hand it is of interest that a closely similar, practically identical syndrome has been described by Stamler (1959) in pregnant rats fed a 'antivitamin E stress diet', rich in oxidized lipids containing unsaturated fatty acids and deficient in vitamin E. The selective sensitivity of the maternal organism to cadmium, specific for the last period of pregnancy, is strictly dependent on the presence of the placenta: hysterectomy or removal of foetuses with placentae before the injection of cadmium salts completely abolished this sensitivity. On the other hand it was possible to induce this 'toxaemia-like syndrome' in rats with surgically removed foetuses when placentae were left in situ (Pařízek et al., 1969a). It is probable that the cadmium effect in the placenta has a decisive role and that the placental lesion induced during this period of pregnancy results secondarily in the specific pathologic changes affecting the maternal organism. It cannot be excluded, of course, that placentae could also sensitize the maternal organism to cadmium by some other mechanisms, changing the reaction of maternal organs to cadmium directly. In any case we can consider this situation as a very good example of how a special physiological condition like pregnancy can profoundly alter the response of the organism to a certain trace element.

The toxicity of cadmium can also be modified by exposure to certain environmental influences. Of particular interest here are the interrelations of cadmium with some other trace elements. The special case of interactions between trace elements in this respect are reviewed in more detail elsewhere (Pařízek 1975).

One of these interrelations is that between cadmium and zinc. Since our first reports on the protective effect of zinc salts against the toxicity of cadmium in the testes (Pařízek 1957), the interrelation between these two atomically homologous elements has been confirmed in many biological systems (Supplee, 1963, Hill et al., 1963, Hennig and Anke 1964, for review and further references see Friberg et al., 1974).

Several reports indicate that increased body burden of cadmium increases the synthesis (increases the level) of metal-binding protein molecules similar to or identical with metallothionein, and that this represents an important protective, detoxifying mechanism (Piscator 1964, Shaikh and Lucis 1971, 1972, Nordberg 1971, Nordberg et al. 1971a, b, Nordberg 1972, Webb 1972a, b, Chen et al., 1974). This could explain why pretreatment with low doses of cadmium salts protects against the toxic effects of a higher, in controls toxic, dose of salts of this metal (Terhaar et al., 1965, Ito and Sawauchi 1965, Gabbiani et al., 1967).

A similar mechanism also seems to play an important role in animals protected against cadmium toxicity by pretreatment with zinc and perhaps with some other (Gabbiani et al., 1967, Yoshikawa 1970) metals. Several laboratories recently reported that pretreatment with zinc salts can increase the synthesis (the level) of low-molecular, metallothionein-like, metal-binding proteins in an analogous way as pretreatment by salts of cadmium (Webb 1972c, Bremner et al., 1973, Davies et al., 1973, Chen et al., 1974). It has been suggested that an increased level of metal-binding proteins, able to sequester cadmium, could well represent the main protective mechanism by which zinc pretreatment alleviates cadmium toxicity (Webb 1972c, Chen et al., 1974). These phenomena are discussed in more detail elsewhere (Pařízek 1975).

The mechanism responsible for the protective effect of another trace element, selenium, seems to be quite different in character. Since the first reports on the protective action of selenite against cadmium-induced testicular damage (Kar et al., 1960, Mason et al., 1964, Mason and Young 1967), administration of small amounts of selenium compounds (selenite or even selenomethionine) was shown to prevent not only all the other effects of cadmium related to reproduction as mentioned above, but cadmium toxicity in general (Pařízek et al., 1969a, 1971a). Rats (Pařízek et al., 1968b) or mice (Gunn et al., 1968) given selenite were shown to survive a very high dose of cadmium compounds lethal in controls. More recently selenite was shown to prevent the cadmium – induced elevation of blood pressure in chronic experiments (Perry and Erlanger 1974).

The protective effect of selenium compounds is not confined to cadmium. Administration of small amounts of selenite was shown to prevent completely the toxic effects of mercuric compounds (Pařízek and Oštadalová 1967, Pařízek et al., 1971). This protective effect is not connected with increased excretion, but on the contrary with increased retention of mercury in the organism (cf. Pařízek et al., 1971).

In animals given cadmium or mercury compounds, selenite or seleno-methionine administration increased the level of these metals in blood and blood plasma by several orders of magnitude (Pařízek et al., 1969b). This characteristic effect was shown to be dependent on the dose of selenium and can be observed even after oral administration of selenium compounds in amounts only slightly higher than those under which selenium operates as an essential trace element (Pařízek et al. 1969b, 1974). A changed binding of the detoxified metal and of selenium could explain why simultaneous exposure to compounds of selenium and mercury (or cadmium) results in altered reactivity and biological availability and

42

in changes in excretion, distribution within the body and transplacental or transmammarian passage of both selenium and the metal involved (Pařízek et al., 1969a, c, 1971b, for further references see Pařízek 1975).

Of particular importance is the fact that the protective effect of selenium can be connected with an increased level of detoxified metal in the blood or even in certain critical organs. The implications resulting from this for the interpretation of results of chemical analyses were discussed previously elsewhere (Pařízek 1972).

A more detailed discussion of interrelation between selenium and cadmium, mercury or certain other metals is given elsewhere (Pařízek et al., 1971, 1974, Pařízek 1975).

However, two points should be stressed here:

i) selenite administration was also shown to protect against the toxicity of mercury given in the form of methylmercury (Ganther et al., 1972, 1973, Stillings et al., 1972, Potter and Matrone 1973, 1974);

ii) very high mercury levels in organs of marine mammals were found to be accompanied by high selenium levels (Koeman et al., 1972, 1973). The very good correlation between the mercury and selenium content and the atomic ratio of 1 : 1 between mercury and selenium found in this study (Koeman et al., 1973) is in very good agreement with the report on the relation between mercury and selenium bound to plasma proteins in experimental animals given mercuric chloride and selenium (Burk et al., 1974). Results suggesting that the interrelation between mercury and selenium also occurs in men exposed to inorganic mercury have recently been reported elsewhere (Kosta et al., 1974).

However, the interrelation between mercury and selenium is very complex in character and does not only have the positive aspect. In our very first paper on mercury and selenium, published several years ago, we already registered a case of surprising death in a group of animals protected against mercury toxicity by selenite administration (Pařízek and Oštádalová 1967). The frequency of this lethality was markedly increased when animals were given mercuric salts not before but few hours after administration of selenite and this effect was particularly pronounced when adult male rats were used instead of females. The character of the symptoms observed, the dependence of the frequence of this syndrome on the sequence of administration as well as certain other factors discussed in detail elsewhere (Pařízek et al., 1971a) seemed to suggest the idea that the most probably explanation here could be a hitherto unknown, markedly potentiating effect of mercuric compounds on the toxicity of dimethylselenide. Our further experiments have confirmed that a very small amount of mercuric compounds can increase the toxicity of this selenium metabolite by several orders of magnitude. Furthermore we observed that this methylated selenium metabolite was more toxic for adult males than for females and that both factors – mercury and sex – had an additive effect (Pařízek et al., 1971a, 1974).

In further experiments we were also able to detect a marked sex-linked difference in the retention of ^{75}Se in animals given ^{75}Se-dimethylselenide (Pařízek et al., 1971a). A series of further experiments (cf. Pařízek et al., 1974), which cannot be discussed in detail here, explained this sex-linked difference in selenium metabolism. We do know now that dimethylselenide is an intermediary metabolite which can be converted into trimethyl-selenonium ions in the organism, the selenium metabolite identified more recently in the urine by Byard (1969) and by Palmer et al. (1969, 1970). We were able to show that the renal excretion of this selenium metabolite is under strict hormonal control of androgens. From the point of view of selenium as an essential trace element and from the point of action of the hormones involved it is of interest that not only androgens but also other anabolic steroids markedly increase the retention of this selenium metabolite in the body. However, we do know as well that this sex-linked difference in selenium metabolism cannot explain the sex-linked difference in the toxicity of dimethylselenide and trimethylselenonium salts. The difference in the retention of these methylated selenium metabolites in the male and female organism appears to be manifested in much younger animals than the sex-linked difference in the toxicity of these selenium compounds. Castration can abolish the sex-linked difference in retention completely. However, it does not affect the sex-linked difference in toxicity of methylated selenium metabolites.

More recently we were able to discover another situation in which the toxicity of methylated selenium metabolites is markedly increased, namely lactation. This increased sensitivity of lactating mothers to the toxicity of dimethylselenide is not connected with increased retention of selenium when given in the form of dimethylselenide or trimethylselenonium ions. The peculiar sensitivity of mothers disappears within 12–24 hrs after artificial weaning (Pařízek et al., 1973, 1974).

Thus we were able to identify three different situations in which the toxicity of methylated selenium metabolites is very markedly increased. Two of these situations are connected with specific physiological charac-teristics – sex and lactation – the third is connected with an important factor of environment. We do not know if and how these three situations in respect to the mechanism of action could have a common denominator. The rate of formation of a toxic metabolite (e.g. monomethylselenide) or the sensitivity to this or to a similar selenium compound could be increased here. In the light of the rapidly growing use of selenium in animal production and in the light of environmental problems connected with mercury we think that this problem is not only interesting but that it also represents an important issue in the solution of which physiology has an important role to play.

SUMMARY

Some physiological aspects of toxic elements are discussed.

Many factors can influence the response of an organism to a toxic element. Some of these factors are: quantity of the toxic element, its chemical form, stage of development of the organism, particular physiological conditions and environmental factors like presence (concentration) of other elements.

The attention is particularly focussed on the toxicity of cadmium salts in relation to animal reproduction and of mercury salts. Some examples of changed toxicity of an element under influence of physiological factors are described: 1. differences in development, sex, pregnancy and hormone-treatment determine the effect of parenterally administered cadmium salts; 2. pretreatment of the animal with zinc salts or low doses of cadmium salts can alleviate cadmium toxicity, probably through increasing the level of a metallothionein-like protein; 3. the protection against cadmium- and mercury-compounds by selenite, but also the potentiating effect of mercury on the toxicity of Se when administered a few hours later than the selenite, are discussed; 4. finally, the potentiating effect of lactation on the toxicity of some selenium compounds is mentioned.

REFERENCES

Bremner, I., N.T. Davies and C.F. Mills, 'The effect of zinc deficiency and food restriction on hepatic zinc proteins in the rat', Biochem. Soc. Trans. 1 (1973) 982.

Burk, R.F., K.A. Foster, P.M. Greenfield and K.W. Kiker, Proc. Soc. Exp. Biol. Med. 145 (1974) 782.

Byard, J.L., 'Trimethyl selenide. A urinary metabolite of selenite', Arch. Biochem. Biophys. 130 (1969) 556.

Chen, R.W., P.A. Wagner, W.G. Hoekstra and H.E. Ganther, 'Affinity labelling studies with cadmium in [109] cadmium-induced testicular injury in rats', J. Reprod. Fert. 38 (1974) 293.

Davies, N.T., I. Bremner and C.F. Mills, 'Studies on the induction of a low-molecular-weight zinc-binding protein in rat liver', Biochem. Soc. Trans. 1 (1973) 985.

Friberg, L., M. Piscator, G.F. Nordberg and Kjellström, '*Cadmium in the Environment*', 2nd ed., (Cleveland, Ohio, 1974).

Gabbiani, G., D. Baic and C. Déziel, *Can. J. Physiol. Pharmacol.* **45** (1967) 443.

Gabbiani, G., M.-G. Badonell, S.M. Mathewson and G.B. Ryan, *Lab. Invest.* **30** (1974) 686.

Ganther, H.E., C. Goudie, M.L. Sunde, M.J. Kopecky, P. Wagner, S.-H. Oh and W.G. Hoekstra, 'Selenium: relation to decreased toxicity of methylmercury added to diets containing tuna', *Sci.* **175** (1972) 1122.

Ganther, H.E., P.A. Wagner, M.L. Sunde and W.G. Hoekstra, 'Protective effects of selenium against heavy metal toxicities', in D.D. Hemphill (ed.), *Trace Substances in Environmental Health* – VI, (University of Missouri Press, Columbia, 1937), 247.

Gunn, S.A., T.C. Gould and W.A.D. Anderson, *Proc. Soc. Exp. Biol. Med.* **128** (1968) 591.

Gunn, S.A. and T.C. Gould, in A.D. Johnson, W.R. Gomes and N.L. Van Demark (eds.), *The Testis*, (Acad. Press, New York 1970), 377.

Hennig, A. and M. Anke, *Arch. Tierernähr.* **14** (1964) 55.

Hill, C.H., G. Matrone, W.L. Payne and C.W. Barber, 'In vivo interactions of cadmium with copper, zinc and iron', *J. Nutr.* **80** (1963) 227.

Kar, A.B., R.P. Das and J.N. Karkun, *Acta Biol. Med. Germ.* **3** (1959) 372.

Kar, A.B., R.F. Das and B. Mukerji, *Proc. Natl. Inst. Sci. India*, Part B. Biol. Sci. **26B** (Suppl.) (1960) 40.

Koeman, J.H., W.H.M. Peeters, C.J. Smit, P.S. Tjioe and J.J.M. de Goeij, 'Persistent chemicals in marine mammals', *TNO-Nieuws* **27** (1972) 570.

Koeman, J.H., W.H.M. Peeters, C.H.M. Koudstaal-Hol, P.S. Tjioe and J.J.M. de Goeij, 'Mercury-selenium correlations in marine mammals', *Nature* **245** (1973) 385.

Kosta, L., A.R. Byrne and V. Zelenko, Presented at the WHO/EPA/CEC Symposium *Recent Advances in the Assessment of the Health Effects of Environmental Pollution*, (Paris, 1974).

Masironi, R.,' Trace elements in relation to cardiovascular diseases: The WHO/IAEA Joint Research Programme', in *Nuclear Activation Techn. in the Life Sc.*, (IAEA, Vienna, 1972), 503.

Mason, K.E., J.A. Brown, J.O. Young and R.R. Nesbit, in O.H. Muth (ed.), *Selenium in Biomedicine*, (Avi Publ. Comp., Westport, Conn., 1967), 383.

Nordberg, G.F., 'Effects of acute and chronic cadmium exposure on the testis of mice', *Environ. Physiol.* **1** (1971) 171.

Nordberg, G.F., M. Piscator and B. Lind, *Acta Pharmacol. Toxicol.* **29** (1971a) 456.

Nordberg, G.F., M. Piscator and M. Nordberg, 'On the distribution of cadmium in blood', *Acta Pharmacol. Toxicol.* **30** (1971b) 289.

Nordberg, G.F., 'Cadmium metabolism and toxicity. Experimental studies on mice with special reference to the use of biological materials as indices of retention and the possible role of metallothionein in transport and detoxification of cadmium', *Environ. Physiol. Biochem.* **2** (1972) 7.

Palmer, I.S., D.D. Fischer, A.W. Halverson and O.E. Olsen, 'Identification of a major selenium excretory product in rat urine', *Biochim. Biophys. Acta* **177** (1969) 336.

Palmer, I.S., R.P. Gunsalus, A.W. Halverson and O.E. Olson, 'Trimethylselenonium ion as a general product from selenium metabolism in the rat', *Biochim. Biophys. Acta* **208** (1970) 260.

Pařízek, J. and Z. Záhoř, 'Effect of cadmium salts on testicular tissue', *Nature* **177** (1956) 1036.

Pařízek, J., 'The destructive effect of cadmium ion on testicular tissue and its prevention

46

by zinc', *J. Endocrinol.* **15** (1957) 56.

Pařízek, J., 'Sterilization of the male by cadmium salts', *J. Reprod. Fert.* **1** (1960) 294.

Pařízek, J., 'Vascular changes at sites of oestrogen biosynthesis produced by parenteral injection of cadmium salts: the destruction of placenta by cadmium salts', *J. Reprod. Fert.* **7** (1964) 263.

Pařízek, J., 'The peculiar toxicity of cadmium during pregnancy – an experimental 'toxaemia of pregnancy' induced by cadmium salts', *J. Reprod. Fert.* **9** (1965) 111.

Pařízek, J. and I. Oštádalová, 'The protective effect of small amounts of selenite in sublimate intoxication', *Experientia* **23** (1967) 142.

Pařízek, J., I. Oštádalová, I Beneš and J. Pitha, 'The effect of a subcutaneous injection of cadmium salts on the ovaries of adult rats in persistent oestrus', *J. Reprod. Fert.* **17** (1968a) 559.

Pařízek, J., I. Oštádalová, I. Beneš and A. Babický, 'Pregnancy and trace elements: the protective effect of compounds of an essential trace element – selenium – against the peculiar toxic effects of cadmium during pregnancy', *J. Reprod. Fert.* **16** (1968b) 507.

Pařízek, J., 'Influence of trace amounts of metals on the reproductive function', in *Yearb. of the Czechoslovak Academy of Sci.* **1967**, (Prague, 1969), 111.

Pařízek, J., I. Beneš, I. Oštádalová, A. Babický, J. Beneš and J. Pitha, 'The effect of selenium on the toxicity and metabolism of cadmium and some other metals', in D. Baltrop and W.L. Burland (eds.), *Miner. Metabolism in Paediatrics*, (Oxford and Edinburgh, Blackwell Scientific Publ., 1969a), 117.

Pařízek, J., I. Beneš, I. Oštádalová, A. Babický, J. Beneš and J. Lener, 'Metabolic interrelations of trace elements. The effect of zinc salts on the survival of rats intoxicated by cadmium', *Physiol. Bohemoslov.* **18** (1969b) 95.

Pařízek J., A. Babický, I. Oštádalová, J. Kalousková and L. Pavlík, 'The effect of selenium compounds on the cross-placental passage of ^{203}Hg', in M.R. Sikov and D.D. Mahlum (eds.), *Radiation Biol. of the Fetal and Juvenile Mammal*, (Oak Ridge, USAEC, 1969c), 137.

Pařízek J., I. Oštádalová, J. Kalouskavá, A. Babický and J. Beneš, 'The detoxifying effects of selenium: Interrelations between compounds of selenium and certain metals', in W. Mertz and W.E. Cornatzer (eds.), *Newer Trace Elements in Nutrition*, (New York, Marcel Dekker, Inc., 1971a), 85.

Pařízek, J., I. Oštádalová, J. Kalousková, A. Babický, L. Pavlík and B. Bíbr, 'Effect of mercuric compounds on the maternal transmission of selenium in the pregnant and lactating rat', *J. Reprod. Fert.* **25** (1971b) 157.

Pařízek, J., 'Toxicological studies involving trace elements', in *Nuclear Activation Techn. in the Life Sci.* (Vienna, IAEA, 1972) 177.

Pařízek, J., J. Kalousková, L. Pavlík, A. Babický and J. Beneš, *Physiol. Bohemoslov.* **22** (1973) 422.

Pařízek, J., J. Kalousková, A. Babický, J. Beneš and L. Pavlík, 'Interaction of selenium with mercury, cadmium, and other toxic metals', in W.G. Hoekstra, J.W. Suttie, H.E. Ganther and W. Mertz (eds.), *Trace Element Metabolism in Animals–2*, (University Park Press, Baltimore, 1974) 119.

Pařízek, J., in G.F. Nordberg (ed.), *Effects and Dose-Response Relationships of Toxic Metals*, (Elsevier, Amsterdam 1975).

Perry, H.M. and M.W. Erlanger, 'Prevention of cadmium-induced hypertension by selenium', *Fed. Proc.* **33** (1974) 357.

Piscator, M., 'On cadmium in normal human kidney together with a report on the isolation of metallothionein from livers of cadmium-exposed rabbits', *Nord. Hyg. Tidskr.* **45** (1964) 76.

47

Potter, S.D. and G. Matrone, 'Effect of selenite on toxicity and retention of dietary methylmercury and mercuric chloride', *Fed. Proc.* **32** (1973) 929.

Potter, S.D. and G. Matrone, 'Effect of selenite on the toxicity of dietary methyl mercury and mercuric chloride in the rat', *J. Nutr.* **104** (1974) 638.

Shaikh, Z.A. and O.J. Lucis, 'Isolation of cadmium-binding proteins', *Experientia* **27** (1971) 1024.

Shaikh, Z.A. and O.J. Lucis, 'Cadmium and zinc binding in mammalian liver and kidneys', *Arch. Environ. Health* **24** (1972) 419.

Stillings, H. Lagally, J. Soares and D. Hiller, in P. Arroyo et al., (eds.), *Abstracts and Communications* 9th International Congress of Nutrition, (Mexico, D.F., 1972) 206.

Supplee, W.C., 'Antagonistic relationship between dietary cadmium and zinc', *Science* **139** (1963) 119.

Terhaar, C.J., E. Vis, R.L. Roudabusch and D.W. Passett, *Toxicol. Appl. Pharmacol.* **7** (1965) 500.

Webb, M., 'Biochemical effects of Cd^{2+}-injury in the rat and mouse testis', *J. Reprod. Fert.* **30** (1972a) 83.

Webb, M., 'Binding of cadmium ions by rat liver and kidney', *Biochem. Pharmacol.* **21** (1972b) 2751.

Webb, M., 'Protection by zinc against cadmium toxicity', *Biochem. Pharmacol.* **21** (1972c) 2767.

Yoshikawa, H., 'Preventive effect of pretreatment with small dose of metals in acute toxicity of metals in mice', *Ind. Health* **8** (1970) 184.

Medical aspects of heavy metals in relations to the environment

I. H. Scheinberg

Earlier during this symposium Hueck (1975) discussed what man is doing to the environment focussing on mercury and copper. He described the biological effects of pollution on two lower organisms, the mussels and gammerids. Parizek (1975) discussed the environment as a multivariable system and discussed interactions of cadmium, selenium, and mercury as models. Finally, Kirchgessner (1975) described specific examples of how the organism reacts to elements which are contaminants. In this paper part of what Hueck said, 'Man contaminates the North Sea with copper and mercury' will be turned around to say, 'Man may be contaminated by these elements'. Copper is the example that will be discussed, principally in physiologic terms. Van den Hamer (1975) will bring the physiology down to the molecular level of interactions between proteins and metals. Hueck (1975) described two kinds of elements: one, of which Hg is an example, is toxic but is not essential. The second kind of element – Cu is an example – is not only potentially toxic (as any element is in sufficient concentration) but is also essential for health and life. Both Cu and Hg can be toxic. The chemical basis of the toxicity of Cu and Hg ions is probably their combination with sulfhydryl groups. Three points constitute the key to their different natures:

	Cu	Hg
1	essential	not essential
2	actively absorbed	no active absorption
3	actively excreted	no active excretion.

As we evolved from our forbears – including the mussels and gammerids – we were faced with an environment of some 92 elements. Some we made use of – constructing oxygen-carrier compounds from iron, or from copper. In reacting to many elements, we were faced with the fact that some we chose to utilize might be either deficient in the environment – and would need to be conserved – or might be present in much higher concentrations than necessary to satisfy our needs and thus might cause toxicity.

Copper is of the latter type: we need it for cytochrome oxidase and for perhaps twenty other proteins and enzymes – but there is still a great excess of it in the environment relative to these needs. So Cu is apparently both absorbed and excreted by active mechanisms present in the gastrointestinal tract.

Mercury, on the other hand, is merely passively absorbed and passively excreted through the gastrointestinal tract, the kidney, and, to a slight degree, through the lungs as mercury vapor.

The nature and sites of Cu and Hg toxicity are remarkably similar, probably because both, as heavy metals, do most of their damage by precipitating or denaturing proteins. Both affect the brain, eye, kidney and the gastrointestinal tract and its associated organs. It are our physiologic responses to each that are significantly different: mercury is a hazard in the environment and copper is not, because if copper is absorbed, at least through the gastrointestinal tract, our active excretory mechanism protects us from toxicity (Table I).

Table I. Comparison of metabolic aspects of copper and mercury

	Copper	*Mercury*
Toxic species	Cu^{2+}	$Hg\uparrow$, Hg^{2+}
Chemical basis of toxicity	Combination with –SH	Combination with –SH for Hg^{2+}
Essentiality	Yes	No
Active absorption	Yes	No
Active excretion	Yes	No
Sites of absorption	GI tract	GI tract
	Burned skin	Skin
	Lungs	Lungs
	Uterine mucosa	
Excretory pathways	GI tract	GI tract
	Kidney (minimal)	Kidney
		Lungs ($Hg\uparrow$) (minimal)
Organs affected	Liver	GI tract
	Brain	Kidney
	Kidney	Brain
	Eye	Eye
		Skin
Industrial hazard	Negligible	Appreciable ($Hg\uparrow$)
Toxic serum conc. of free ionic species	$> 10 \ \mu g/100$ ml	$> 20 \ \mu g/100$ ml
Therapeutic agents	Penicillamine	Penicillamine
	B.A.L.	N-acetyl penicillamine
		B.A.L.

$Hg\uparrow$ = mercury as vapor

Table II. Effects of copper deficiency in animals

Disturbances in iron metabolism
Diminution in phospholipid synthesis
Defective osteoblastic activity
Abnormalities of keratin and pigment

The essentiality of Cu is demonstrated by the data summarized in Tables I and II. Copper deficiency occurs in man in Menkes' disease probably as a result of an hereditary defect in the active mechanism of absorption and transport (Danks et al., 1973). Whereas the livers of normal infants contain more than 100 μg Cu/g dry weight, in children with Menkes' disease there may be only 10 or 20 μg/g (Danks et al., 1972).

Table III. Mammalian copper proteins*

Protein	Isolated from	
	Species	*Organ or tissue*
Albocuprein I	Man	Brain
Albocuprein II	Man	Brain
Ceruloplasmin	Numerous, incl. Man	Plasma
Cytochrome c oxidase	Numerous	Heart, liver, etc.
3,4-Dihydroxyphenylethylamine β-hydroxylase	Cattle	Adrenals
Dopamine β-hydroxylase	Cattle	Adrenals
Ferroxidase II	Man	Serum
Hepatomitochondrocuprein	Man, Cattle	Liver
L-6D (metallothionein)	Man, Cattle	Liver
Lysyl oxidase	Chicken	Cartilage
Mitochondrial monoamine oxidase	Man, Rat, Cattle	Liver, brain
Pink copper protein	Man	Erythrocytes
Plasma/serum monoamine oxidase	Man, Rabbit, Pig	Plasma/serum
Superoxide dismutase (cytocuprein)		
Cerebrocuprein	Man	Brain
Erythrocuprein	Man	Erythrocytes
Hemocuprein	Man	Blood
Hepatocuprein	Man	Liver
Tryptophan-2, 3-dioxygenase	Rat	Liver
Tyrosinase	Man	Skin, eye

* From I. Sternlieb and I. H. Scheinberg, 'Human copper metabolism in relation to the environment', in *Report of the Panel on Copper*, Committee on Medical and Biologic Effects of Environmental Pollutants. National Academy of Sciences, (1975) (in preparation).

These children exhibit mental retardation, kinky hair and deficient concentrations of cytochrome oxidase, and ultimately die due to the lack of copper. Genetically normal infants, with respect to copper absorption, who have diarrhea and receive a diet low in copper can also develop and manifest signs of copper deficiency (Cordano et al., 1966). The existence of specific mammalian copper-proteins (Table III) also attests to the essentiality of the metal.

Finally, one can demonstrate the essentiality of Cu in man if deficiency is induced pharmacologically. Hypogeusia, or diminished taste acuity, develops in a fairly large proportion of individuals with normal Cu metabolism when penicillamine, a drug that chelates and promotes excretion of copper in the urine, is administered to them. The disturbance is promptly corrected when copper is administered orally, even if penicillamine therapy is continued (Henkin et al., 1967).

The other side of the penny is the toxicity of copper. This is seen in sheep – principally as hemolytic crises, or liver disease – if they feed on pastures containing an excess of Cu (Underwood, 1971). Chickens may also show hemolytic anemia as a sign of copper toxicity. In fish, mice and rats Cu toxicity can be readily induced experimentally, so there is no doubt that Cu is a potential poison, as Hueck (1975) has stated. But the extreme rarity of copper toxicity in man is due to an active inherited mechanism of excretion, congenitally absent in patients with Wilson's disease (Scheinberg and Sternlieb, 1965). This mechanism, the obverse of the active absorptive one whose absence appears to underlie Menkes' disease, is remarkably effective. In a good Dutch diet there are perhaps 5 mg of copper every day – and we need little more than a few tenths of a milligram to keep us in balance and supply our needs. So we have to rid ourselves of 4 to 5 mg per day, and patients with Wilson's disease, lacking this mechanism, accumulate copper. There is no race we know of where this abnormal 'Wilson's disease gene' does not occur: it is a very democratic, autosomal, recessive gene. Each parent possesses one abnormal 'Wilson's disease gene' and one normal 'Cu balance gene'. In accord with Mendelian laws of inheritance, one-quarter of the children of such parents will have a pair of the abnormal genes – and Wilson's disease develops only in these abnormal homozygotes.

About one in 200,000 individuals has Wilson's disease. This means that about one in 200 individuals is heterozygous for the Wilson's disease gene – is, in other words, a carrier in whom the disease is never seen. In a non-consanguineous marriage each partner has a one in 200 chance of being heterozygous, and the chance that two unrelated partners are both heterozygous is, therefore, 1 in 200^2, or 1 in 40,000. Since only one out of four children will inherit a pair of Wilson's disease genes, the frequency of homozygous children will be 1 in 4 × 40,000, or roughly 1 in 200,000 – neglecting the effects of consanguinity. Thus, our diploid character protects the vast majority of us from Wilson's disease.

Fig. 1. Crystals of ceruloplasmin, prepared by fractionation of human plasma by Anatol G. Morell. Photomicrograph was made by Irmin Sternlieb and Milton Kurtz.

About 95 % of patients with Wilson's disease are characterized by a deficiency of the blue plasma protein, ceruloplasmin, isolated, described and named 25 years ago, in Sweden, by Laurell and Holmberg (Fig. 1). A quarter of these patients appears to have no detectable ceruloplasmin at all (Table IV). Frieden (1970), claims that ceruloplasmin is a ferroxidase

Table IV. Concentrations of ceruloplasmin in serum of symptomatic patients with Wilson's disease

Ceruloplasmin mg/100 ml	Patients No.	Per Cent
0 – 1	60	22.0
1 – 4.9	79	28.9
5 – 9.9	67	24.6
10 – 14.9	35	12.8
15 – 19.9	18	6.6
≥ 20	14	5.1
Total	273	100.0

(Normal: 20–40 mg/100 ml)

essential to the normal metabolism of iron. However, even those patients with Wilson's disease with no detectable ceruloplasmin never seem to exhibit any unexplained abnormality in iron metabolism or anemia. Ceruloplasmin may not be essential to iron metabolism in man, or a second ferroxidase may exist.

Visible copper deposition occurs in Wilson's disease in the periphery of the cornea, forming the Kayser-Fleischer ring. Although this ring does not seem to have any adverse effects, the excessive accumulation of copper in the liver, brain, and kidneys most certainly does, producing severe disorders of the liver and central nervous system which, untreated, are always fatal.

A number of copper chelating agents have been proposed as therapeutic agents to rid these patients of excess copper. Penicillamine – discovered and first used by Walshe in England (Walshe 1956) – has proved to be the most successful. This compound, β, β-dimethylcysteine, has the capacity to combine with Cu, and to promote its urinary excretion. Within a year of beginning therapy with penicillamine, the indices of liver dysfunction and the neurologic or psychiatric disturbances improve or disappear completely in the large majority of patients with Wilson's disease.

One patient,* a man who was 23 years old when admitted to our hospital, showed a remarkable improvement – from bed-ridden incapacity to virtual normality – within three months after treatment was begun. He exemplifies what can be done by removing this accumulated, environmental pollutant provided one does not wait too long. In the same month as his admission, a woman of the same age – and like the man mis-diagnosed as schizophrenic, and given a number of electro-shock treatments – was also admitted. But, in contrast to the man, she is still unable to talk or swallow and her body remains rigid and spastic. It is clear that the manifestations of copper poisoning in her are irreversible. Thus, the clinician must not delay in making, or excluding, the diagnosis of Wilson's disease and must promptly institute treatment, if the disorder is present, to avoid permanent damage, or death (Sternlieb and Scheinberg, 1968; Sternlieb and Scheinberg, 1964).

The time course of the development of symptoms in Wilson's disease is shown in Fig. 2: no patient has been noted to be overtly ill before the age of six years, and several patients have first developed symptoms in the fifth decade of life. The period in which most patients manifest illness is between 15 and 20 years, with a median age of about 15.

When this disease was first described by Wilson in 1911, it was always progressive and fatal. Therapeutic success made us wonder if one could make the diagnosis biochemically before Kayser-Fleischer rings, liver or central nervous system disease occurred. Treatment of such an asymptomatic individual might forestall any clinical disorder (Sternlieb and Scheinberg, 1960). From our study of patients with overt disease, we formulated a pair of necessary and sufficient criteria for the diagnosis of Wilson's disease in any individual, sick or well: a deficiency of ceruloplasmin (< 20 mg%) *and* an excess of hepatic copper ($> 250\,\mu g/g$ dry liver). Based largely on these criteria, we have, over the past 15 years, diagnosed 70 asymptomatic patients. (Yet, in histologic sections of the liver of 61 of them, in whom biopsy was performed, 54 showed definite pathologic structural abnormalities). All 70 patients were treated with D-penicillamine for a total of 535 patient-years of therapy. During this period, two developed Kayser-Fleischer rings, one a slight tremor, and one psychiatric manifestations. Of 11 similar asymptomatic patients, who for various reasons were not treated with D-penicillamine, all became overtly ill with Wilson's disease, and five of them have died (Sternlieb and Scheinberg, 1968; Sternlieb, 1972; Sternlieb and Scheinberg, 1973).

* During his lecture Prof. Scheinberg showed an 11 minute movie, illustrating the incapacitating symptoms of Wilson's disease and the dramatic improvement achieved with therapy in the case of this patient, 'The Intensive Treatment of Wilson's Disease', produced by I.H. Scheinberg and I. Sternlieb, (1974).

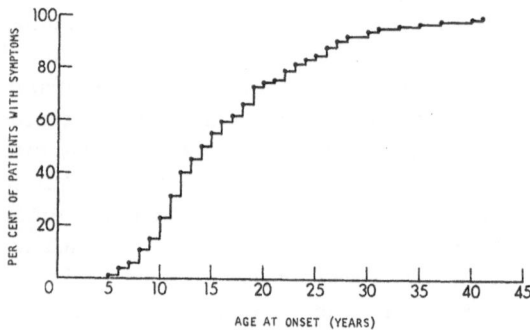

Fig. 2. Cumulative percentage of 121 patients versus age at onset of overt symptoms of Wilson's disease.

We are learning of an increasing number of diseases in which an hereditary defect underlies the disorder. We are still far from the type of genetic engineering that can repair or replace an abnormal gene. But, as Wilson's disease exemplifies, we can sometimes at least treat the phenotypic effects of the abnormal gene by treating the environment – that of a patient to rid him of the effects of a pollutant. By environment in these cases we refer to Claude Bernard's 'milieu intérieur' – but the process is closely related to interactions of environment and man.

The genetics of hereditary copper poisoning may be somewhat more complex than indicated above. Pauling originally described sickle cell hemoglobin as the first genetically abnormal form of a protein, but since then more than a hundred different molecular forms of human hemoglobin alone have been reported, each the consequence of a different abnormal gene. Similarly, Wilson's disease may be associated with at least two abnormal alleles of the normal 'copper-balance' gene. In a computerized study of over 100 families with Wilson's disease (Scheinberg and Sternlieb, unpubl. observ.) we found seven families in which there were two or three siblings with symptoms of the disease that were predominantly neurological in all the siblings (Table V). Further, in any one of these families the siblings all fell sick within three years of each other, despite the great age spread in onset shown in Fig. 2. There were another five families in which the disease was predominantlyhepatic, and all the siblings of each family fell ill within 1.7 years of each other. There were three families in which the disease was discordant in siblings both with respect to clinical form and age at onset.

56

Table V. Onset of Wilson's disease in families with 2 or 3 symptomatic siblings

Mode of onset	No. of families	Mean difference in age at onset within sibships
Concordant (CNS)	7*	3.3
Concordant (hepatic)	5*	1.7
Discordant	3*	5.2

* 1 family with 3 symptomatic siblings.

One family is particularly interesting because two brothers married two unrelated women. One brother had three children, in all of whom the disease developed. The other brother had three children with the disease and two who were either carriers or homozygously normal. All three children of the first brother had an hepatic onset very close to nine years of age. The children of the other brother all had a central nervous form of onset, and all had it close to twelve years of age. It seems very unlikely that both of these unrelated wives had the same abnormal gene as the obviously identical abnormal gene of the two brothers. These data suggest that there are quite possibly two abnormal alleles of the normal copper balance gene that can give rise to Wilson's disease – as shown in Fig. 3 – which lead to somewhat different abnormalities in copper metabolism.

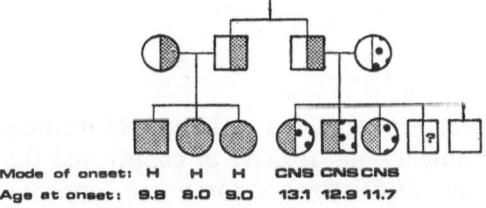

Mode of onset: H   H   H   CNS CNS CNS
Age at onset: 9.8 8.0 9.0   13.1 12.9 11.7

Fig. 3. Pedigree of six cousins with Wilson's disease. H, hepatic onset; CNS, central nervous system onset.

We have seen that copper is potentially toxic to man and that most of us avoid this toxicity by an inherited mechanism. It seems difficult – perhaps impossible – to defeat this protective mechanism. Copper miners we have studied uniformly had normal levels of ceruloplasmin and hepatic copper and showed no evidence of copper toxicity whatever despite twenty or more years of exposure in an open pit mine, eating their lunches in an atmosphere of dust containing 2 % of copper (Scheinberg and Sternlieb, 1969).

Yet, it is barely possible that this mechanism may not be capable of protecting man from copper introduced parenterally. In the U.S., the Dutch discovery of chronic hemodialysis for renal disease is carried out clinically through cellophane membranes fabricated with copper to improve their permeability characteristics. Individuals dialyzed twice a week or more for one to two years can accumulate as much as ten to twenty times as much copper in their livers as do normal people (Manzler et al., 1970; Barbour et al., 1971). Has evolution given us a protective mechanism only against copper that we eat? The evolutionary process never faced the introduction of copper parenterally – and the mechanism may be less effective against such copper. May some symptoms of Wilson's syndrome develop ultimately in chronic hemodialysis patients? And is this relevant to the new intrauterine copper containing contraceptive devices? These consist of about 100 mg of fine copper wire – wound around plastic, and inserted in the uterus. Are we completely safe in putting such amounts of copper into a woman for ten to thirty years in light of the demonstration of its absorption: in a rat intrauterine ^{64}Cu can be demonstrated in the liver and serum within 18 hours after insertion in the uterus (Okereke et al., 1972). Evolution has never faced that situation before!

SUMMARY

The toxic effects of copper in man are compared with those of mercury. The evidence for both the essentiality and toxicity of copper and the toxicity of mercury is summarized. Deficiency and toxicity of copper are both rare occurrences in man, the latter being virtually limited to the genetic disorder Wilson's disease in which the normal inherited mechanism for excreting copper is defective. The physiological buffering that so effectively prevents deficiency and toxicity of copper in man is probably a consequence of the essentiality of the metal for health and life – and is in contrast to mercury for which evolution has found no physiological function.

REFERENCES

Barbour, B.H., M. Bischel and D.E. Abrams, 'Copper accumulation in patients undergoing chronic hemodialysis. The role of Cuprophan', *Nephron* 8 (1971) 455.
Cordano, A., R.P. Placko and G.G. Graham, 'Hypocupremia and neutropenia in copper deficiency', *Blood* 28 (1966) 280.
Danks, D.M., E. Cartwright, B.J. Stevens and R.R.W. Townley, 'Menkes' kinky hair disease: Further definition of the defect in copper transport', *Science* 179 (1973) 1140.
Danks, D.M., B.J. Stevens, P.E. Campbell, J.M. Gillespie, J. Walker-Smith, J. Blomfield and B. Turner, 'Menkes' kinky-hair syndrome', *Lancet* i (1972) 1100.
Frieden, E. 'Ceruloplasmin, a link between copper and iron metabolism', *Nutr. Rev.* 28 (1970) 87.
Henkin, R.I., H.R. Keiser, I.A. Jaffe, I. Sternlieb and I.H. Scheinberg, 'Decreased taste sensitivity after D-penicillamine, reversed by copper administration', *Lancet* ii (1967) 1268.
Hueck, H.J. 'Contamination of the environment by some elements', *This symposium*, (1975).
Kirchgessner, M., 'Resorption and metabolism of essential elements', *This symposium*, (1975).
Manzler, A.D. and A.W. Schreiner, 'Copper-induced acute hemolytic anemia. A new complication of hemodialysis', *Ann. Int. Med.* 73 (1970) 409.
Okereke, T., I. Sternlieb, A.G. Morell, and I.H. Scheinberg, 'Systemic absorption of intrauterine copper', *Science* 179 (1972) 358.
Parizek, J., 'Physiology of toxic elements', *This symposium*, (1975).
Scheinberg, I.H. and I. Sternlieb, 'Wilson's disease', *Ann. Rev. Med.* 16 (1965) 119.
Scheinberg, I.H. and I. Sternlieb, 'Metabolism of trace elements', in P.K. Bondy (ed.), *Duncan's Diseases of Metabolism*, vol. 2, 6th ed., (Philadelphia, Saunders, 1969), p. 1321.
Sternlieb, I. 'Evolution of the hepatic lesion in Wilson's disease (hepatolenticular degeneration)', in H. Popper, F. Schaffner (eds.), *Progress in Liver Diseases*, vol. 4, (Grune and Stratton, New York, 1972), p. 511.
Sternlieb, I. and I.H. Scheinberg, 'Penicillamine therapy in hepatolenticular degeneration', *J. Amer. Med. Ass.* 189 (1964) 748.
Sternlieb, I. and I.H. Scheinberg, 'Prevention of Wilson's disease in asymptomatic patients', *New Eng. J. Med.* 278 (1968) 352.
Sternlieb, I. and I.H. Scheinberg, 'Reassessment of prophylactic therapy of patients with asymptomatic Wilson's disease', *Proc. 3rd Intern. Symp. on Wilson's Disease*, (Paris, Sept. 1973).
Underwood, E.J., *Trace Elements in Human and Animal Nutrition*, 3rd ed. (Acad. Press, New York, 1971).
Van den Hamer, C.J.A., 'Metal-protein interactions on the molecular level'. *This symposium*, (1975).
Walshe, J.M., 'Wilson's disease'. *New oral therapy*, Lancet i (1956) 25.

DISCUSSION

Dr. L.N. Went, Leiden

What method would you suggest to detect the presence of homozygotes before any clinical symptoms have occurred since most of the future patients will be the first affected child of a couple?

Prof. I.H. Scheinberg

I suggest the determination of ceruloplasmin in serum, between the age of six months and six years, in every child, with quantitative copper determination in a biopsy sample of liver in all children with unexplained deficiency of ceruplasmin. Most efficiently, this should be done as part of a screening program for a number of hereditary disorders.

Prof. M. Kirchgessner, Munich

Have you ever injected a patient having Wilson's disease with ceruloplasmin, with or without copper?

Prof. I.H. Scheinberg

Yes, about twenty years, ago, both ceruloplasmin and apoceruloplasmin were given to patients. The first is not only of no value, but is harmful since it increases the copper content of the body. Apoceruloplasmin disappears rapidly from the plasma and apparently is catabolized without combining with Cu.

Prof. M. Kirchgessner, Munich

Is there a possibility that penicillamine and Cu, already present in the G.I. tract, will form a complex which is not absorbed? With oxytetracycline we note a better absorption of Cu but the tetracycline-Cu complex, on the other hand, is not absorbed.

Prof. I.H. Scheinberg

Probably not. Penicillamine is given 30 to 60 minutes before meals in order to avoid its association with dietary copper most of which is, of course, not absorbed, either in normal subjects or in those with Wilson's disease.

Dr. G.N. Tytgat, Amsterdam

How does ceruloplasmin protect against Wilson's disease?

Prof. Scheinberg

This is not known, but, perhaps it does so by binding copper in a molecule of about 150,000 daltons, thereby making it unable to diffuse out of the vascular compartment and into the brain, eyes, etc. Or, perhaps, it provides a pathway of excretion of excess copper, via catabolism of ceruloplasmin, and consequent excretion of its copper.

Dr. H.G. Eyk, Rotterdam

Has the work described in Science (G. W. Evans et al., *Science 151* (1973) 1175) on a liver protein in Wilson's disease with a four times higher binding constant than the protein from normal livers been confirmed?

Prof. I.H. Scheinberg

No. That work is, I believe, incorrect since the data used in the Scatchard plot to calculate the binding constant were not obtained at equilibrium. The calculations of the paper are, therefore, invalid.

Metal-protein interactions on the molecular level

C. J. A. Van den Hamer

Even a first look at the information available about the physiology of trace metals shows the complexity and diversity of the subject. Kirchgessner and coworkers (Kirchgessner, 1975; Grassmann, 1973) showed the dependance on a number of factors not only of the resorption of trace metals from the gastrointestinal tract but in some instances also of their metabolic efficiency – the extent to which they can be utilized in the metalloenzymes etc. of the tissues. Parizek (1975) pointed out the mutual influencing of trace metals. How widely different trace metals can behave in the mammalian body is also illustrated in Table I and Table II in which some of the results obtained on intravenous injection of these metals in the rat under fixed conditions are summarized.

Table I. Distribution of intravenously injected metals in the rat. Values in % of the dose per g or ml 2 h after an intravenous injection of 10 µg of the metal ion in 0.05 M acetate buffer pH 5.6, 0.15 M NaCl.
Rats: male Wistar 200 ± 10 g.

	Cd^{2+}	Cr^{3+}	Cu^{2+}	Fe^{3+}	Hg^{2+}	Mn^{2+}	Pb^{2+}	Zn^{2+}
Blood	0.010	0.67	0.34	2.55	1.62	0.008	1.47	0.30
Plasma	0.015	1.20	0.57	5.36	0.70	0.007	0.030	0.25
Liver	7.94	8.84	3.81	1.97	1.32	3.45	1.38	2.70
Kidneys	1.96	0.30	6.22	1.37	16.0	2.96	11.56	5.0
Bone*	n.d.	0.50	n.d.	2.83	3.18	0.657	n.d.	0.67
Brain	0.023	0.014	0.026	0.09	0.027	0.054	0.031	0.096
Fur	0.21	0.069	0.34	0.10	0.26	0.245	0.082	0.24
Muscle	0.041	0.049	0.065	0.13	0.09	0.062	0.023	0.14
Pancreas+	1.5	0.09	0.31	0.31	0.17	1.86	0.40	2.77
Spleen	n.d.	6.98	n.d.	2.10	0.48	0.61	n.d.	2.16

* femur + tibia of hindleg
+ calculated per total organ
n.d. = not determined

Table II. Excretion by the rat of intravenously injected metals. Values in % of the dose. Bile was collected during 2½ h by canulation under light nembutal anesthesia. At the end of this period the content of the small intestine was collected by rinsing with a 1% NaCl solution. Other experimental conditions as in Table I.

	Cd^{2+}	Cr^{3+}	Cu^{2+}	Fe^{3+}	Hg^{2+}	Mn^{2+}	Na^+	Pb^{2+}	Zn^{2+}
Bile canulation 2½ h									
in collected bile	2.6	0.06	15.9	0.05	0.19	16.7	3.8	n.d.	0.06
in content small intestine	n.d.	0.29	5.1	0.24	2.4	4.0	2.2	n.d.	0.2
Metabolic cage 96 h									
in collected feces	n.d.	n.d.	63.2	8.5	16.1	41.2	n.d.	21.6	30.0
in collected urine	n.d.	n.d.	5.4	1.2	2.5	0.4	n.d.	14.7	0.2

n.d. = not determined

The mammalian body can handle a number of trace metals usually quite efficiently; it ensures itself of the right amount by maintaining the proper balance between intake and excretion, although sometimes these mechanisms fail in case of extreme conditions or a genetic defect (Scheinberg, 1975). Another failure in the natural defence, e.g. against mercury, is due to recent sudden changes in our natural surroundings, brought about by man himself, which the organism is unable to cope with since the time for adaption through the evolutionary process was too short (Hueck, 1975). Yet, no matter how diverse the responses of the mammalian body to the various trace metals are, these metals will almost invariably be found in the organism bound to proteins. It is this aspect that will be stressed in the following discussion of some routes of copper and, to a lesser extent, of other trace metals through the human body.

A schematic and simplified picture of the main routes of copper through the human body is given in Fig. 1, which is a composite of various pieces of information from the literature. As this diagram shows, we daily ingest with our food some 2–5 mg of copper, part of which is resorbed from the small intestine and enters the blood. Here, in the blood, this freshly arrived copper is bound to albumin and is transported in this form to the various tissues. A substantial part will go to the liver which plays a central role in maintaining homeostasis. Some of the liver copper will be incorporated in the plasma protein ceruloplasmin during its synthesis in the liver. Because the concentration of this protein in the plasma is constant, an equal amount must be degraded per day, which is also a task of the liver.

Since the total copper content of the body does not vary appreciably, the daily intake and excretion must be equal. The balance is maintained by excretion of the excess of copper, mainly through the bile, while smaller quantities are reported to be lost through leakage of copper-carrying proteins – ceruloplasmin and albumin – directly into the lumen of the gastrointestinal tract. The excretion through the urine is minor, but losses through sweat, although numbers are not yet available, possibly play a significant role (Hohnadel et al., 1973) as seems to be the case for zinc (WHO, 1973).

In the following a few aspects of the above scheme will be discussed in some detail and a comparison will be made between copper and some of the other trace metals.

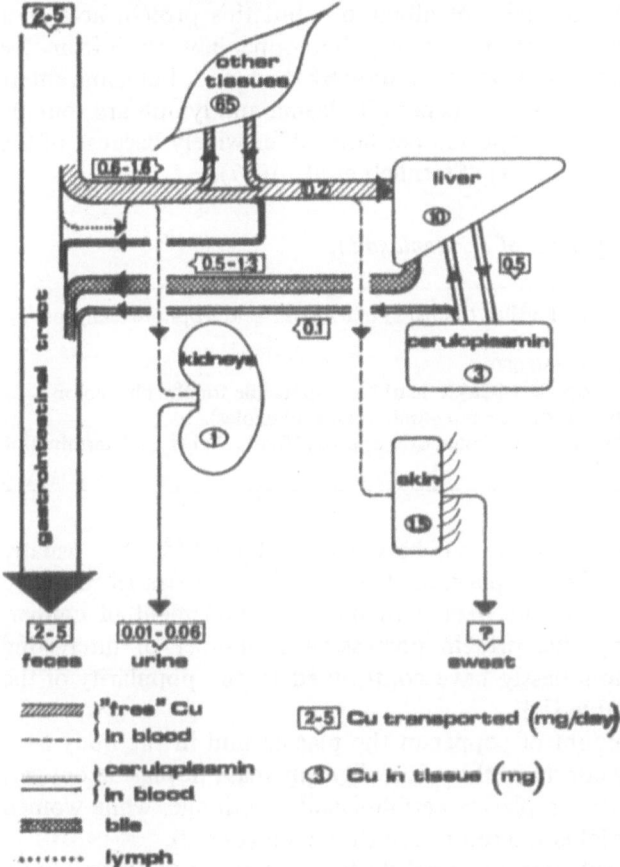

Fig. 1. Diagram of pathways of dietary copper through the human body. Ceruloplasmin expressed in terms of its Cu-content.

As shown in Fig. 1., after resorption the copper is transported by the blood to the various tissues. It is carried loosely bound to albumin. In this form it is usually referred to as 'free Cu' or 'direct reacting Cu', terms referring to the fact that the copper in this form is much less tightly bound and therefore reacts more readily with reagents than copper in ceruloplasmin. According to Breslow (1964) the copper is probably bound to a terminal amino group and two peptide bond nitrogens. The albumin has such a size and shape that it can readily enter the extravascular spaces where it can pick up or give off copper, depending on the local conditions.

Only 5% of the plasma copper is bound to albumin – or about 1 atom of copper per 1000 molecules of albumin – but this protein acts as a very efficient vehicle for copper. Thus, the copper absorbed from the gastrointestinal tract is transported mainly by the blood, but some enters the lymph. Although the amounts per ml plasma and lymph are roughly the same, the amounts transported per hour differ widely because of the much slower flow of the latter (Sternlieb et al., 1967).

Table III. Some properties of ceruloplasmin.

1. Plasma protein – normal value (humans) 0.3 mg/ml - electrophoretically an α_2-globulin.
2. Molecular weight about 150,000.
3. Contains 8 atoms of copper – $4Cu^{2+}$, $4Cu^{1+}$ – responsible for the blue color and the oxidase activity (substrates: poly-amines and -phenoles).
4. Contains 8% carbohydrate, distributed over approx. 10 side chains, each terminating in sialic acid.

Most of the copper in plasma, however, – about 95% – is usually present in the form of ceruloplasmin. This protein is, as described above (Fig. 1), synthesized by the liver with its full complement of copper. It is a very pretty, blue protein possessing a number of interesting properties which doubtlessly have contributed to the popularity of the study of copper (Table III).

Although the amount of copper in the plasma and in the body as a whole is kept fairly constant, there is a slight increase in plasma copper, because of an increase in plasma ceruloplasmin, with age, while women show a somewhat higher concentration than men (Fig. 2).

The copper in ceruloplasmin is tightly bound but it can reversibly be removed forming the so called apo-ceruloplasmin (Morell and Scheinberg, 1958). The protein can even be taken apart to its Cu-free subunits

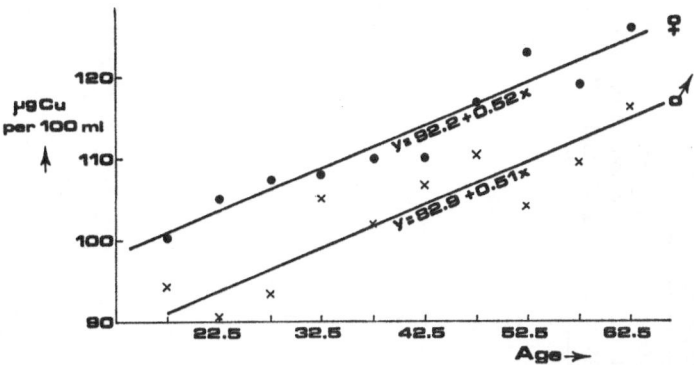

Fig. 2. Copper concentration in human serum as function of the age of the donor. Each point represents the average of 5 samples. [G. Buyze, C.J.A. Van den Hamer and M.C.M. van der Heyden; unpubl. results].

and reassembled again with good yield (Freeman, 1973). But forms with less than their full complement of copper have also been reported in Cu-deficient rats and in patients with Wilson's disease (Holtzman et al., 1970a, b; Van den Hamer et al., to be published). Finally, the copper atoms of ceruloplasmin can, under reducing conditions, exchange with external copper atoms, but in vivo no such exchange – and therefore no role in copper transport – can be demonstrated (Sternlieb et al., 1961). However, in this connection it is interesting that Marceau and Aspin (1973) notrd that when ceruloplasmin is catabolized in the liver part of its copper – but not plasma 'free copper' – becomes an intrinsic part of mitochondiral cytochrome c oxidase. A role in the oxidation of Fe^{2+} to Fe^{3+}, a step needed for the incorporation of iron in transferrin, has been proposed as a ceruloplasmin function (Frieden, 1973; but see also Scheinberg, 1975).

Several other trace metals are, like Cu, found to be transported bound to albumin, e.g., Zn, Mn and Ni. Iron on the other hand has its own specific carrier, transferrin, which can carry two atoms of ferrous iron per molecule. And, like Cu is incorporated in ceruloplasmin, similar plasma proteins are found containing other trace metals. The properties of some of these are summarized in Table IV. Both the Zn containing α_2-macroglobulin and nickeloplasmin show esterase activity, but otherwise very little is known about the role of these proteins which should be classified as metalloproteins since the metal forms an integral part of it. McBean et al. (1974) found that in some pathologic conditions α_2-

Table IV. Comparison of properties of some plasma metalloproteins.

Metal	Protein	Mol. weight	mg protein/ ml plasma	atoms metal/ mol. protein	Fraction of metal in plasma bound to protein	Reference
Cu	Ceruloplasmin	150,000	0.3	8	0.95	see text
Ni	Nickeloplasmin	700,000	?	1	0.45	Nomoto et al., 1973
Zn	Zn α_2-macroglobulin	840,000	2.4	3.3 – 7.9	0.35	Parisi and Vallee, 1970

macroglobulin was elevated while the plasma Zn tended to be low, but that in 6 cases of primary Zn deficiency the concentration of α_2-macroglobulin was several times higher than normal.

Trace metals are also present in plasma in the form of enzymes. However, in this form they represent very small amounts of metal indeed – plasma alkaline phosphatase for instance accounts for less than 1 ng of zinc per ml plasma. They mostly originate from breakdown of tissue cells.

More important from point of view of trace metals are the amino acids in the plasma. Small amounts of copper (Neumann and Sass-Kortsak, 1967; Hallman et al., 1971) and zinc (Hallman et al., 1971) are bound to histidine and cystine and to histidine and cysteine, resp. Though these are quantitatively also minor forms, they play an important role since these amino acids are thought to be the carriers of the trace metals across membranes like those of the liver cells.

Let us now turn our attention to the liver. When a rat is intravenously injected with a solution of ^{64}Cu, much of this radionuclide will go very fast to the liver from where it gradually disappears again. To understand the processes involved, one has to look more closely at the liver cells. It was shown that roughly 2/3 of the liver copper is present in the soluble fraction of the cells, the cytosol, while the balance is distributed over the various subcellular particles: nuclei, mitochondria, etc. (Feldman et al., 1972). The same is true for the copper that enters the liver as the result of an intravenous injection, providing the dose was not too large (Fig. 3-A). It seems therefore that, in order to learn more about the copper metabolism, one has to focus on the liver cytosol.

Homogenizing the liver of a rat injected before with ^{64}Cu, preparing a particle free cytosol by centrifugation and subjecting this cytosol to gelfiltration, one can show that the ^{64}Cu is mainly associated with two materials with apparent molecular weights of approx. 10,000 and 35,000, indicated in Fig. 3 as Fr-I and Fr-II, resp. The ratio of ^{64}Cu associated with these two materials clearly depends on the time since the injection and on the dose, as a comparison of the two parts of Fig. 3 shows. This figure also illustrates how increasing the dose results in an increased involvement of the particulate fraction of the liver.

The results of the gelfiltration experiments, combined with those of determination of the specific activity of the copper in the cytosol fractions, suggest that copper after entering the liver is first attached to a protein of molecular weight 10,000, that is then transferred to (or the first protein is converted into) a second protein with molecular weight of 35,000 and that it is finally excreted via the bile into the lumen of the gastrointestinal

69

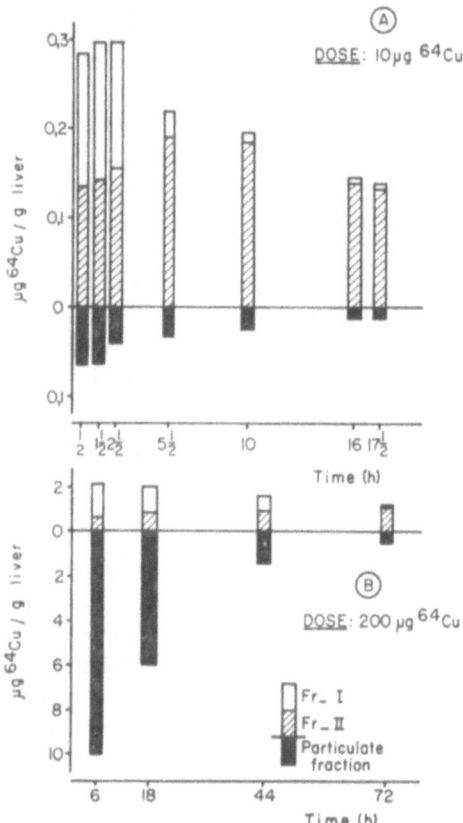

Fig. 3. Distribution of ^{64}Cu over fractions of the rat liver as function of time since its intravenous injection. Fr-I and Fr-II: Cu-binding proteins with molecular weights of 10,000 and 35,000, resp. Dose: as indicated. [C.J.A. Van den Hamer et al., to be publ.].

tract (Fig. 4). The involvement of the particulate fraction must be mainly regarded as a side effect. This effect will become exaggerated when a large dose of copper has been given, conditions under which apparently the normal mechanisms fail to cope efficiently with the influx of copper. Similar accumulations of Cu in the liver subcellular particles are also found when rats are intraperitoneally injected for several weeks with copper (Feldmann et al., 1972) and in the case of patients with Wilson's disease (Sternlieb et al., 1973).

70

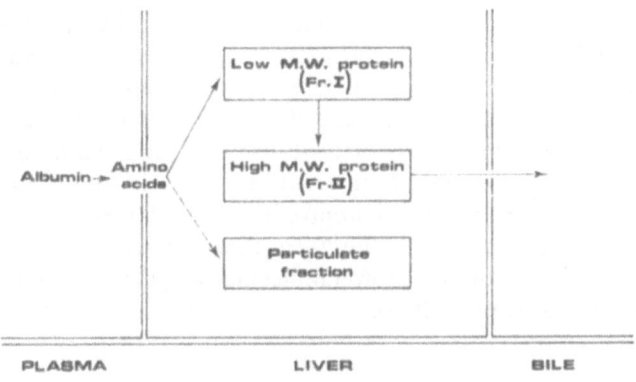

Fig. 4. Diagram of the relation between some Cu-binding fractions in the liver of the rat.

One of the copper binding proteins of the liver described above, the 10,000 molecular weight material, is the same as the protein called L-6D by Morell et al., (1961). The metal in this protein is bound to sulfhydryl groups, much like Cd and Zn in the protein metallothionein, a metal binding protein isolated by Kägi and Vallee (1960) from horse kidney. A protein similar to metallothionein has been found in the livers of rats subjected to intraperitoneal injections of Cd (Winge and Rajagopalan, 1972). Besides these and similar metal binding proteins, a number of proteins with the same molecular weight has been described which bind in the liver such substances as bilirubin, sulfobromophthalein, etc. (Levi et al., 1969; Kaplowitz et al., 1973). Since quite often little more has been done than noting their apparent molecular weight, it seems prudent for the moment not to conclude too quickly to identity or non-identity of such proteins.

With an antiserum, prepared by immunizing rabbits with a purified preparation of the liver copper protein of molecular weight 35,000, it could be shown that other tissues of the rat contained a protein that was similar to, if not identical with, the corresponding liver protein. However, no clear correlation was observed in the various tissues between the concentration of this protein, the copper content and the ability to bind intravenously injected ^{64}Cu. Neither has the relation of this protein with the enzyme superoxide dismutase yet been clarified.

Many of the observations made with the other trace metals are in brought lines similar to, but in detail quite different from those with

71

copper. For instance, most of the liver zinc and iron is, like copper, found in the cytosol but most of the manganese, magnesium and calcium on the other hand can be found in the nuclei fraction (Thiers and Vallee, 1957). Further analysis with gelifitration usually reveals also in these cases more than one metal binder in the liver cytosol (Fig. 5). Much of a dose of ^{65}Zn, for instance, when intravenously injected into a rat, is found after a short time connected with a protein of approximately 100,000 molecular weight. During the following hours the ^{65}Zn slowly decreases in this fraction and appears in fractions with lower molecular weight (Stortenbeek and Van den Hamer, to be published).

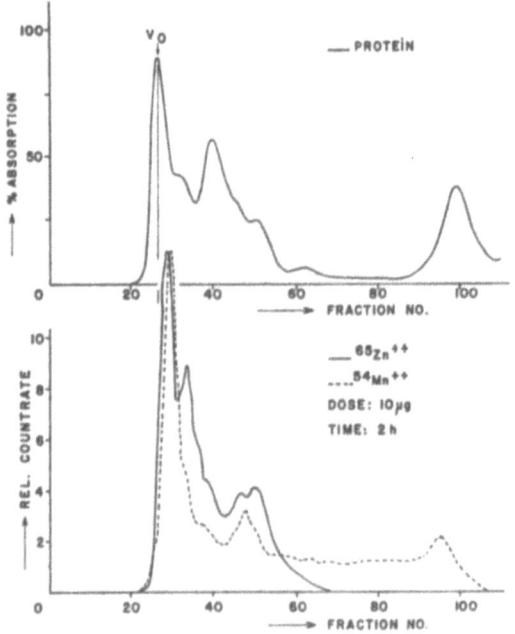

Fig. 5. Gelfiltration of liver cytosol prepared from rats, intravenously injected with ^{65}Zn or ^{54}Mn. In the lower curve the distributions of the two radionuclides are compared. The upper curve shows the distribution of the protein (% absorption at 280 nm; same distribution for both experiments). Other conditions: rats: 200 g – decapitated 2 h after injection; dose: 10 μg; gelfiltration: column Sephadex G-75 (1.8 × 80 cm) – eluted with 0.05 M Tris buffer pH 7.0, 0.7% NaCl – fractions of 1.25 ml.

As shown in Fig. 4 the two copper binding proteins probably play a role in the homeostasis of the copper by facilitating the excretion of excess copper via the bile. This picture is supported by the observation that in case of Wilson's disease – in which copper accumulates among others in the liver (Scheinberg, 1974) – the copper content of the bile is lower than normal while little or no 35,000 molecular weight copper binding protein (Fr-II) is found in the liver, suggesting that the primary difficulty of this disease lays in an inadequate transfer of copper from Fr-I to Fr-II (Sternlieb et al., 1973). On the other hand it is known that zinc is mainly excreted via the pancreas and through the duodenal wall, while little or no zinc reaches the gastrointestinal tract via the bile (Barrowman et al., 1973; Gunasekera et al., 1973; see also Table II). Neither the form in which the zinc is transported to be duodenal wall nor the function of the zinc liver proteins is yet known.

From these examples it is clear that the differences between binding proteins in the liver cytosol for the various metals are more than superficial and that much more knowledge of the properties and physiology of these and other metal binding proteins is needed for a better understanding of the behavior of the trace metals.

SUMMARY

Pathways of dietary copper through the body and the proteins to which it is bound at the various stages, e.g., albumin and ceruloplasmin in the plasma and two proteins of molecular weight of 10,000 and 35,000, resp. in the liver cytosol, are described. A scheme involving these liver proteins is proposed for the excretion of copper via the bile.

Similar proteins are mentioned in connection with other trace metals. More knowledge of such proteins is needed for better understanding of the behaviour of trace metals.

REFERENCES

Barrowman, J.A., R. Bonnett and P.J. Bray, 'Biliary excretion of zinc in rats', *Biochem. Soc. Trans.* **1** (1973) 988.
Breslow, E., 'Comparison of cupric ion-binding sites in myoglobin derivatives and serum albumin', *J. Biol. Chem.* **239** (1964) 3252.
Feldmann, G., C. Abramowitz, H. Sarmini and F. Rousselet, 'Repartition du cuivre dans les fractions subcellulaires hépatiques après intoxication subaiguë par le cuivre chez le rat', *Biol. Gastro-Entérol.* **5** (1972) 37.
Freeman, S. and E. Daniel, 'Dissociation and reconstitution of human ceruloplasmin', *Biochem.* **12** (1973) 4806.

Frieden, E., 'The ferrous to ferric cycles in iron metabolism', *Nutr. Rev.* **31** (1973) 41.
Grassmann, E., *Kupfer im tierischen Organismus, sein Stoffwechsel und seine Wechsel-wirkungen mit Ionen und Liganden der Nährung*, (Technische Universität, München-Weihenstephan, 1973).
Gunasekera, S.W., L.J. King and D.V. Parke, 'The excretion of [65Zn] zinc in the rat', *Biochem. Soc. Trans.* **1** (1973) 900.
Hallmann, P.S., D.D. Perrin and A.E. Watt, 'The computed distribution of copper (II) and zinc (II) ions among seventeen amino acids present in human blood plasma', *Biochem. J.* **121** (1971) 549.
Hohnadel, D.C., F.W. Sunderman Jr, M.W. Nechay and M.D. McNeely, 'Atomic absorption spectrophotometry of nickel, copper, zinc, and lead in sweat collected from healthy subjects during sauna bathing', *Clin. Chem.* **19** (1973) 1288.
Holtzman, N.A. and B.M. Gaumnitz, 'Identification of an apoceruloplasmin-like substance in the plasma of copper-deficient rats', *J. Biol. Chem.* **245** (1970) 2350.
Holtzman, N.A. and P. Charache, A. Cordano and G.G. Graham, 'Distribution of serum copper in copper deficiency', *Hopkins Med. J.* **126** (1970) 34.
Hueck, H.J., 'Contamination of the environment by some elements', *This symposium*, (1975).
Kägi, J.H.R. and B.L. Vallee, 'Metallothionein: a cadmium- and zinc-containing protein from equine renal cortex', *J. Biol. Chem.* **235** (1960) 3460.
Kaplowitz, N., I.W. Percy-Robb and N.B. Javitt, 'Role of hepatic anion-binding protein in bromsulphthalein conjugation', *J. Exp. Med.* **138** (1973) 483.
Kirchgessner, M., 'Resorption and metabolism of essential elements', *This symposium*, (1975).
Levi, A.J., Z. Gatmaitan and I.M. Arias, 'Two hepatic cytoplasmic protein fractions, Y and Z, and their possible role in the hepatic uptake of bilirubin, sulfobromophthalein, and other anions', *J. Clin. Invest.* **48** (1969) 2156.
Marceau, N. and N. Aspin, 'The intracellular distribution of the radiocopper derived from ceruloplasmin and from albumin', *Biochim. Biophys. Acta* **293** (1973) 338.
McBean, L.D., J.C. Smith Jr., B.H. Berne and J.A. Halsted, 'Serum zinc and alpha2-macroglobulin concentration in myocardial infarction, decubitus ulcer, multiple myeloma, prostatic carcinoma, Down's syndrome and nephrotic syndrome', *Clin. Chim.* **50** (1974) 43.
Morell, A.G. and I.H. Scheinberg, 'Preparation of an apoprotein from ceruloplasmin by reversible dissociation of copper', *Science* **127** (1958) 588.
Morell, A.G., J.R. Shapiro and I.H. Scheinberg, 'Copper binding protein of human liver', in J.M. Walshe, J.N. Cumings (eds.), *Wilson's Disease*, (Oxford, Blackwell Scientific Publ., 1961), 36.
Neumann, P.Z. and A. Sass-Kortsak, 'The state of copper in human serum: evidence for an amino acid-bound fraction', *J. Clin. Invest.* **46** (1967) 646.
Nomoto, S., M.I. Decsy, J.R. Murphy and F.W. Sunderman Jr., 'Isolation of 63Ni-labelled nickeloplasmin from rabbit serum', *Biochem. Med.* **8** (1973) 171.
Parisi, A.F. and B.L. Vallee, 'Isolation of a zinc α_2-macroglobulin from human serum', *Biochem.* **9** (1970) 2421.
Parizek, J., 'Physiology of toxic elements', *This symposium*, (1975).
Scheinberg, I.H., 'Medical aspects of heavy metals', *This symposium*, (1975).
Sternlieb, I., A.G. Morell, W.D. Tucker, M.W. Green and I.H. Scheinberg, 'The incorporation of copper into ceruloplasmin in vivo: studies with copper64 and copper67', *J. Clin. Invest.* **40** (1961) 1834.
Sternlieb, I., C.J.A. Van den Hamer and S. Alpert, 'Role of intestinal lymphatics in

copper absorption', *Nature* **216** (1967) 824.

Sternlieb, I., C.J.A. Van den Hamer, A.G. Morell, S. Alpert, G. Gregoriadis and I.H. Scheinberg, 'Lysosomal defect of hepatic copper excretion in Wilson's disease (hepatolenticular degeneration)', *Gastroenterology* **64** (1973) 99.

Thiers, R.E. and B.L. Vallee, 'Distribution of metals in subcellular fractions of rat liver', *J. Biol. Chem.* **226** (1957) 911.

Winge, D.R. and K.V. Rajagopalan, 'Purification and some properties of Cd-binding protein from rat liver', *Arch. Biochem. Biophys.* **153** (1972) 755.

'Trace elements in human nutrition', *Wld. Hlth. Org. tech. Rep. Ser.*, (1973), No. 532.

DISCUSSION

Drs. H. Nederbragt, Utrecht

1. Ceruloplasmin was described as a non-transport protein. How does this agree with the finding of Marceau and Aspin (*Biochim. Biophys. Acta 293* (1973) 338, 351) that ceruloplasmin is involved in transfer of copper to cytocuprein (mol. weight 35.000)?
2. What is the site of the ceruloplasmin synthesis? Where does its copper come from?
3. What are the characteristics of the copper that is present in increased concentration in patients with Wilson's disease?

Dr. C.J.A. Van den Hamer

1. Ceruloplasmin is, like most plasma proteins, catabolized in the liver. As a result of this process its copper becomes free and is handled like copper from albumin. According to Marceau and Aspin some of the ceruloplasmin copper becomes an intrinsic part of liver cytochrome c oxidase.
2. Ceruloplasmin is synthesized by and only by the liver. We could not detect any incorportaion of ^{64}Cu in ceruloplasmin in hepatectomized rats. Its copper seems to be derived from a small copper pool (not shown in Fig. 4) because the specific activity of the ^{64}Cu in ceruloplasmin synthesized shortly after an intravenous injection of this nuclide in rats resembles more that of the injection than that of the copper proteins of the liver cytosol.
3. In the liver of Wilson's patients the copper is connected with the protein of mol. weight of 10,000 (Fr-I in Fig. 4) and with the subcellular particles elevated.

Prof. I.H. Scheinberg, New York

Metallothionein and L-6D may not be the same protein but L-6D and the protein isolated by Evans et al., (*Science 151* (1973) 1175) are probably the same since the methods of isolation are similar.

Dr. H.G. van Eyck, Rotterdam

By which methods are metals removed from plasma proteins in vivo?

Dr. C.J.A. Van den Hamer

Copper carried by albumin is probably taken over by amino acids and transported in this form into the cells. Ceruloplasmin on the other hand enters as intact protein the liver cells and is there catabolized. For zinc probably similar mechanisms play a role.